CELL BIOLOGY: MEIOSIS AND MITOSIS

Edited by
Kristy June

a
ArclerPress

Cell Biology: Meiosis and Mitosis

Edited by Kristy June

Edition 2018

ArclerPress

Published by
Arcler Press LLC
708 3rd Avenue, 6th Floor
New York
NY 10017
United States of America
www.arclerpress.com

ISBN: 978-1-68094-535-5 (Hardback)

Printed in United States of America

Library of Congress Control Number: 2017933122

About the Editor

Kristy June

Kristy June is a Doctor of Medicine who also masters copyediting and proofreading scientific and academic research papers. She finished Bachelor of Science in Biology Major in Ecology and Entomology in the University of the Philippines Mindanao as her undergraduate degree where she started engaging herself into writing, editing and proofreading articles and books for publications. She is currently working as both a Medical Doctor and an Editor

Table of Contents

3 THE PROCESS OF MITOSIS

4 MATURATION PROMOTING FACTOR

5 MITOTIC BOOKMARKING

6 ROLE OF A MITOGEN

7 MITOTIC INDEX

8 MITOTIC CATASTROPHE AND CARCINOGENESIS

List of Abbreviations

AM	Adrenomedullin
APC	Anaphase, Promoting Complex
CTD	Carboxyl Terminal Domain
CDKs	Cyclin-Dependent Kinases
DS	Down Syndrome
EGF	Epidermal Growth Factor
EPO	Erythropoietin
FBS	Foetal Bovine Somatotrophin
GDF9	Growth Differentiation Factor-9
HGF	Hepatocyte Growth Factor
HDGF	Hepatoma-Derived Growth Factor
LIF	Leukemia Inhibitory Factor
MAPK	Mitogen-Activated Protein Kinase
MDS	Myelodysplastic Syndrome
NGF	Nerve Growth Factor
PGF	Placental Growth Factor
PBD	Polar Body Diagnosis
PGD or PIGD	Pre-Implantation Genetic Diagnosis
ROS	Reactive Oxygen Species
TPO	Thrombopoietin
TMD	Transient Myeloproliferative Disorder
TL	Transient Leukemia

VEGF	Vascular Endothelial Growth Factor
SAC	Spindle Assembly Checkpoint

Preface

Cells divide and reproduce in two ways: mitosis and meiosis. Mitosis is a process of cell division that results in two genetically identical daughter cells, developing from a single parent cell. Meiosis, on the other hand, is the division of a germ cell, involving two fissions of the nucleus and giving rise to four gametes, or sex cells, each possessing half the number of chromosomes of the original cell. Mitosis is used by single-celled organisms to reproduce; it is also used for the organic growth of tissues, fibers, and membranes. Meiosis is found in sexual reproduction of organisms. The male and female sex cells (i.e., egg and sperm) are the end result of meiosis; they combine to create new, genetically different offspring. Both mitosis and meiosis are associated with cytokinesis. The end result of both are daughter cells produced from a parent cell. The fundamental sequence of events in mitosis is the same as in meiosis (in meiosis it happens twice). Both processes include the breakdown of the nuclear membrane, the separation of genetic material into two groups, followed by cell division and the reformation of the nuclear membrane in each cells. The processes differ in two fundamental. Meiosis has two rounds of genetic separation and cellular division while mitosis only has one of each. In meiosis homologous chromosomes separate leading to daughter cells that are not genetically identical. In mitosis the daughter cells are identical to the parent as well as to each other. One of the benefits of sexual reproduction is the diversity it produces within a population. That variety is a direct product of meiosis. Every sex cell made from meiosis has a unique combination of chromosomes. This means that no two sperm or egg cells are genetically identical. Every fertilization event produces new combinations of traits. This is why siblings share DNA with parents and each other, but are not identical to one another. Mitosis is important for sexual reproduction indirectly. It allows the sexually reproducing organism to grow and develop from a single cell into a sexually mature individual. This allows organisms to continue to reproduce through the generations.

Content Coverage

Chapter One is anticipated to present the introduction of cell division by which a parental cell gives rise to two daughter cells. The process involves both nuclear division and cytokinesis and can either produce two equal cells (symmetric cell division) or two cells with different cellular fates (asymmetric cell division).

Chapter Two presents the process of meiosis that reduces the chromosome number by half, creating four haploid cells, each genetically distinct from the parent cell that gave rise to them. This process occurs in all sexually reproducing single-celled and multicellular eukaryotes, including animals, plants, and fungi.

Chapter Three highlights the process of mitosis in which a eukaryotic cell nucleus splits in two, followed by division of the parent cell into two daughter cells. This process ensures that each daughter cell will contain one exact copy of the parent cell DNA.

Chapter Four highlights about the maturation promoting factor (MPF) that regulates the passage of a cell from the G2 growth phase to the M phase. It is also known as the G2 checkpoint, and ensures that DNA replication during the S phase did not produce any mistakes. It also makes sure that the cell grew enough during the G2 stage.

Chapter Five reveals about mitotic bookmarking that refers to a potential mechanism of transmission of gene expression programs through cell division. During mitosis, gene transcription is silenced and most transcription factors are removed from chromatin.

Chapter Seven focuses on the role of a mitogen that triggers the process of mitosis, or cell division. Frequently, mitogens are proteins – compounds consisting of amino acid molecules that facilitate biochemical reactions in organisms. Mitogens are often encountered in the field of immunology due to their importance in inducing signal transduction pathways.

Chapter Seven emphasizes the purpose of the mitotic index is to measure cellular proliferation. The mitotic index is an important prognostic factor predicting both overall survival and response to chemotherapy in most types of cancer. It may lose much of its predictive value for elderly populations.

Chapter Eight discusses about the mechanism of mitotic catastrophe of delayed mitosis-linked cell death, a sequence of events resulting from premature or inappropriate entry of cells into mitosis that can be caused by chemical or physical stresses. Mitotic catastrophe is unrelated to programmed cell death or apoptosis and is observed in cells lacking functional apoptotic pathways.

CHAPTER 1

Cell Division: An Introduction

Chapter Contents

Cell division is the process by which a parental cell gives rise to two daughter cells. The process involves both nuclear division and cytokinesis and can either produce two equal cells (symmetric cell division) or two cells with different cellular fates (asymmetric cell division).

INTRODUCTION

Cell division is the process by which a parent cell divides into two or more daughter cells.Cell division usually occurs as part of a larger cell cycle. In eukaryotes, there are two distinct types of cell division: a vegetative division, whereby each daughter cell is genetically identical to the parent cell (mitosis), and a reproductive cell division, whereby the number of chromosomes in the daughter cells is reduced by half to produce haploid gametes (meiosis). Meiosis results in twelve haploid daughter cells by undergoing one round of DNA replication followed by two divisions: homologous chromosomes are separated in the first division, and sister chromatids are separated in the second division. Both of these cell division cycles are used in sexually reproducing organisms at some point in their life cycle, and both are believed to be present in the last eukaryotic common ancestor.

Prokaryotes also undergo a vegetative cell division known as binary fission, where their genetic material is segregated equally into two daughter cells. All cell divisions, regardless of organism, are preceded by a single round of DNA replication.

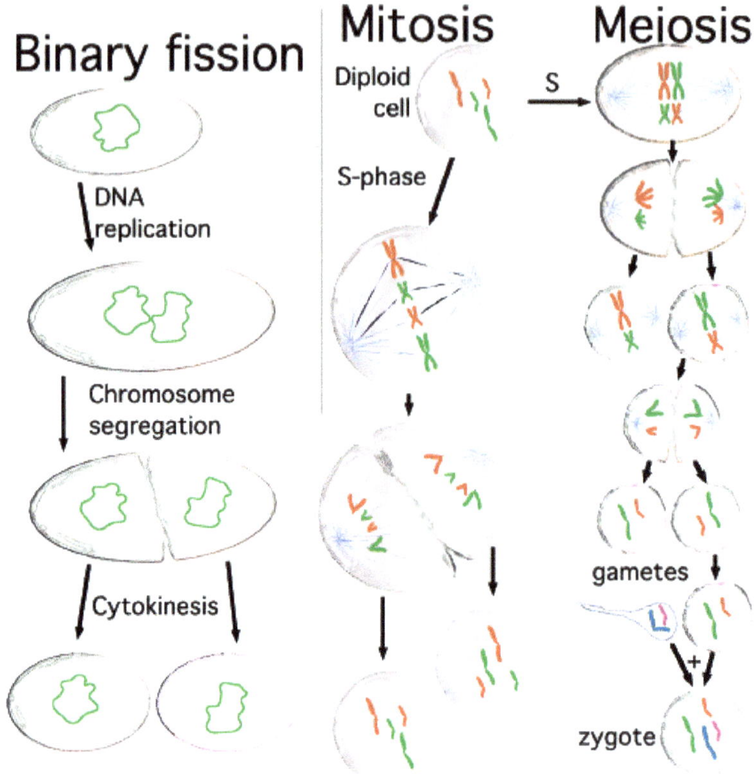

Three types of cell division.

For simple unicellular organisms such as the amoeba, one cell division is equivalent to reproduction – an entire new organism is created. On a larger scale, mitotic cell division can create progeny from multicellular organisms, such as plants that grow from cuttings. Mitotic cell division also enables sexually reproducing organisms to develop from the one-celled zygote, which itself was produced by meiotic cell division from gametes. After growth, cell division by mitosis allows for continual construction and repair of the organism. The human body experiences about 10 quadrillion cell divisions in a lifetime.

The primary concern of cell division is the maintenance of the original cell's genome. Before division can occur, the genomic information that is stored in chromosomes must be replicated, and the duplicated genome must be separated cleanly between cells. A great deal of cellular infrastructure is involved in keeping genomic information consistent between generations.

VARIANTS

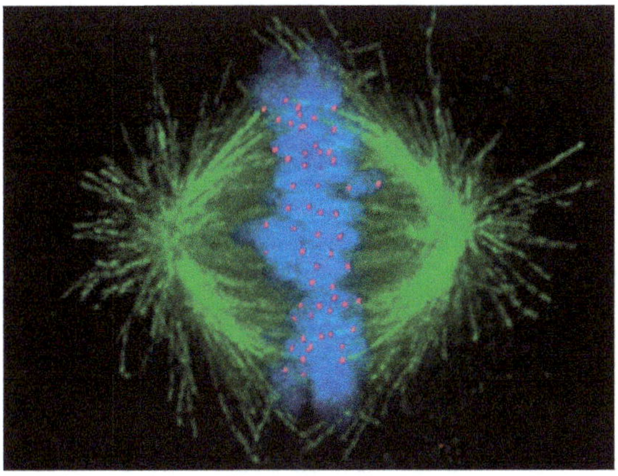

Image of the mitotic spindle in a human cell showing microtubules in green, chromosomes (DNA) in blue, and kinetochores in red.

Cells are broadly classified into two main categories: simple, non-nucleated prokaryotic cells, and complex, nucleated eukaryotic cells. Owing to their structural differences, eukaryotic and prokaryotic cells do not divide in the same way. Also, the pattern of cell division that transforms eukaryotic stem cells into gametes (sperm cells in males or egg cells in females), termed meiosis, is different from that of the division of somatic cells in the body.

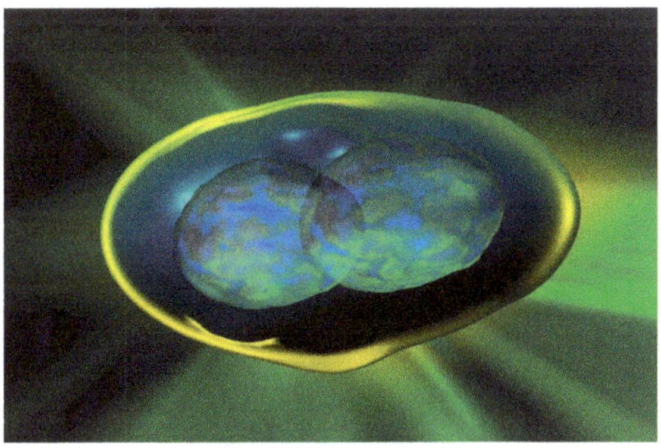

Cell division over 42 hours. The cells were directly imaged in the cell culture vessel, using non-invasive quantitative phase contrast time-lapse microscopy.

DEGRADATION

Multicellular organisms replace worn-out cells through cell division. In some animals, however, cell division eventually halts. In humans this occurs, on average, after 52 divisions, known as the Hayflick limit. The cell is then referred to as senescent. Cells stop dividing because the telomeres, protective bits of DNA on the end of a chromosome required for replication, shorten with each copy, eventually being consumed. Cancer cells, on the other hand, are not thought to degrade in this way, if at all. An enzyme called telomerase, present in large quantities in cancerous cells, rebuilds the telomeres, allowing division to continue indefinitely.

BIOLOGICAL LIFE CYCLE

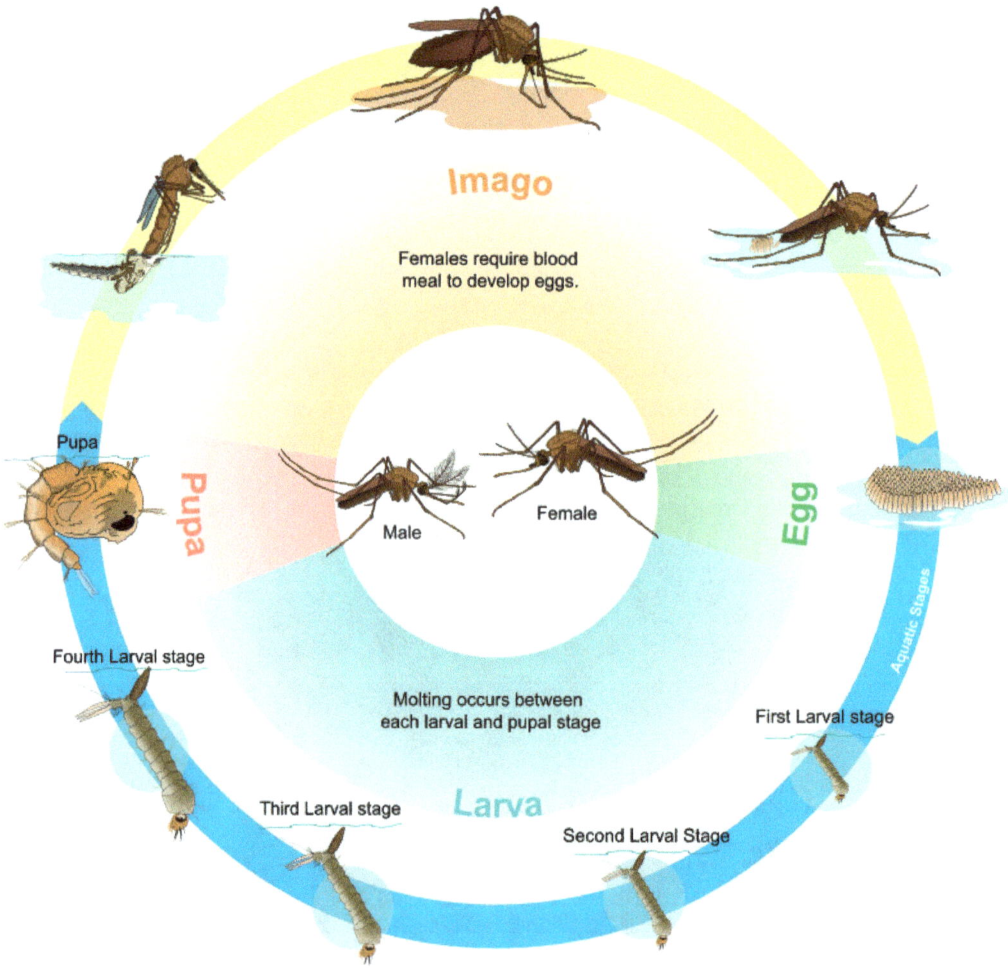

Life cycle of a mosquito. An adult (imago) lays eggs which develop through several stages to adulthood. Reproduction completes and perpetuates the cycle.

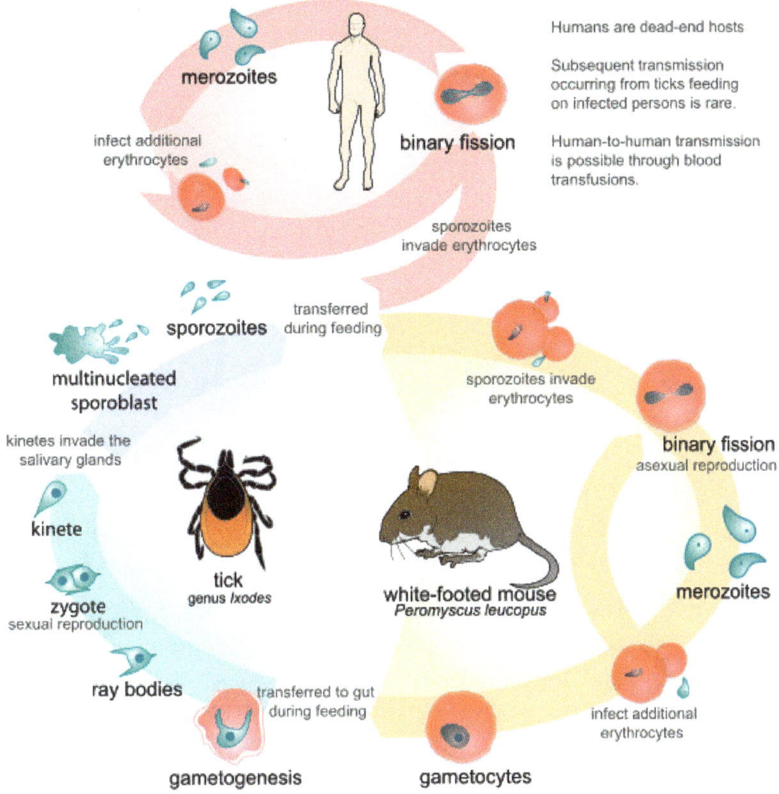

Life cycle of the single-celled parasite *Babesia*.

In biology, a **biological life cycle** (or just **life cycle** when the biological context is clear) is a series of changes in form that an organism undergoes, returning to the starting state. "The concept is closely related to those of the life history, development and ontogeny, but differs from them in stressing renewal.Transitions of form may involve growth, asexual reproduction, and/or sexual reproduction.

In some organisms, different generations of the species succeed each other during the life cycle. For plants and many algae, there are two multicellular stages, and the life cycle is referred to as alternation of generations. The term **life history** is often used, particularly for organisms such as the red algae which have three multicellular stages (or more), rather than two.

Life cycles that include sexual reproduction involve alternating haploid (n) and diploid ($2n$) stages, i.e., a change of ploidy is involved. To return from a diploid stage to a haploid stage, meiosis must occur. In regard to changes of ploidy, there are 3 types of cycles:

- ***haplontic*** life cycle — the haploid stage is multicellular and the diploid stage is a single cell, meiosis is "zygotic".
- ***diplontic*** life cycle — the diploid stage is multicellular and haploid gametes are formed, meiosis is "gametic.

- ***haplodiplontic*** life cycle (also referred to as *diplohaplontic*, *diplobiontic*, or *dibiontic* life cycle) — multicellular diploid and haploid stages occur, meiosis is "sporic.

The cycles differ in when mitosis (growth) occurs. **Zygotic meiosis** and **gametic meiosis** have one mitotic stage: mitosis occurs during the *n* phase in zygotic meiosis and during the 2*n* phase in gametic meiosis. Therefore, zygotic and gametic meiosis are collectively termed **haplobiontic** (single mitotic phase, not to be confused with haplontic). **Sporic meiosis**, on the other hand, has mitosis in two stages, both the diploid and haploid stages, termed **diplobiontic** (not to be confused with diplontic).

Discovery

The study of reproduction and development in organisms was carried out by many botanists and zoologists.

Wilhelm Hofmeister demonstrated that alternation of generations is a feature that unites plants, and published this result in 1851 (see plant sexuality).

Some terms (haplobiont and diplobiont) used for the description of life cycles were proposed initially for algae by Nils Svedelius, and then became used for other organisms. Other terms (autogamy and gamontogamy) used in protist life cycles were introduced by Karl Gottlieb Grell. The description of the complex life cycles of various organisms contributed to the disproof of the ideas of spontaneous generation in the 1840s and 1850s.

Haplontic life Cycle

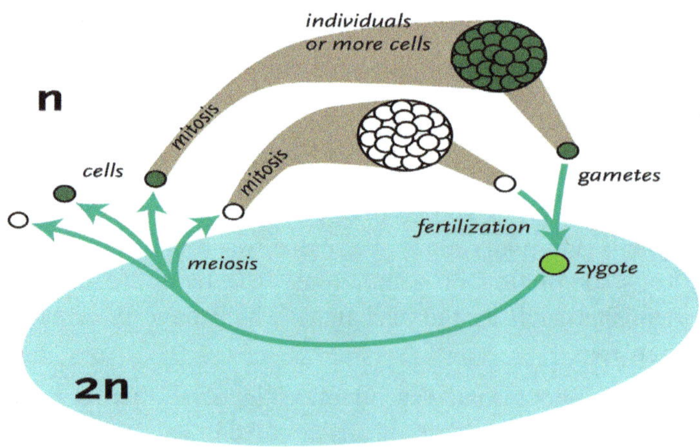

Zygotic meiosis.

A **zygotic meiosis** is a meiosis of a zygote immediately after karyogamy, which is the fusion of two cell nuclei. This way, the organism ends its diploid phase and produces several haploid cells. These cells divide mitotically to form either larger, multicellular individuals, or more haploid cells. Two opposite types of gametes (e.g., male and female) from these individuals or cells fuse to become a zygote.

In the whole cycle, zygotes are the only diploid cell; mitosis occurs only in the haploid phase.

The individuals or cells as a result of mitosis are **haplonts**, hence this life cycle is also called **haplontic life cycle**. Haplonts are:

- In archaeplastidans: some green algae (e.g., *Chlamydomonas*, *Zygnema*, *Chara*)
- In stramenopiles: some golden algae
- In alveolates: many dinoflagellates eg.Ceratium,Gymnodinium, some apicomplexans (e.g., *Plasmodium*)
- In rhizarians: some euglyphids, ascetosporeans
- In excavates: some parabasalids
- In amoebozoans: *Dictyostelium*
- In opisthokonts: most fungi (some chytrids, zygomycetes, some ascomycetes, basidiomycetes)

Diplontic life Cycle

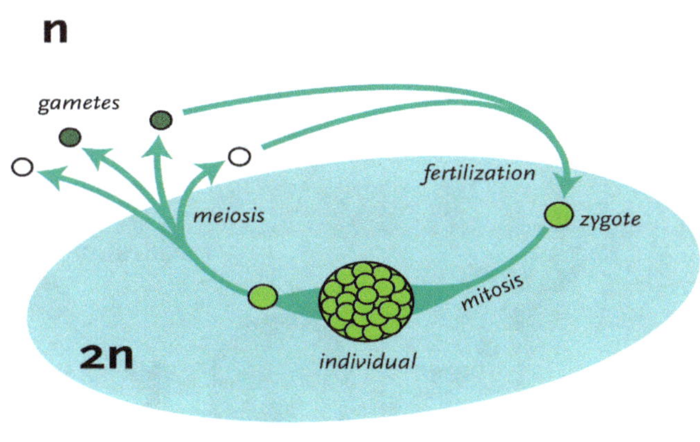

Gametic meiosis.

In **gametic meiosis**, instead of immediately dividing *meiotically* to produce haploid cells, the zygote divides *mitotically* to produce a multicellular diploid individual or a group of more unicellular diploid cells. Cells from the diploid individuals then undergo meiosis to produce haploid cells or gametes. Haploid cells may divide again (by mitosis) to form more haploid cells, as in many yeasts, but the haploid phase is not the predominant life cycle phase. In most diplonts, mitosis occurs only in the diploid phase, i.e. gametes usually form quickly and fuse to produce diploid zygotes.

In the whole cycle, gametes are usually the only haploid cells, and mitosis usually occurs only in the diploid phase.

The diploid multicellular individual is a **diplont**, hence a gametic meiosis is also called a **diplontic life cycle**. Diplonts are:

- In archaeplastidans: some green algae (e.g., *Cladophora glomerata*, *Acetabularia*)
- In stramenopiles: some brown algae (the Fucales, however, their life cycle can also be interpreted as strongly heteromorphic-diplohaplontic, with a highly reduced gametophyte phase, as in the flowering plants), some xanthophytes (e.g., *Vaucheria*), most diatoms, some oomycetes (e.g., *Saprolegnia*, *Plasmopara viticola*), opalines, some "heliozoans" (e.g., *Actinophrys*, *Actinosphaerium*)
- In alveolates: ciliates
- In excavates: some parabasalids
- In opisthokonts: animals, some fungi (e.g., some ascomycetes)

Haplodiplontic Life Cycle

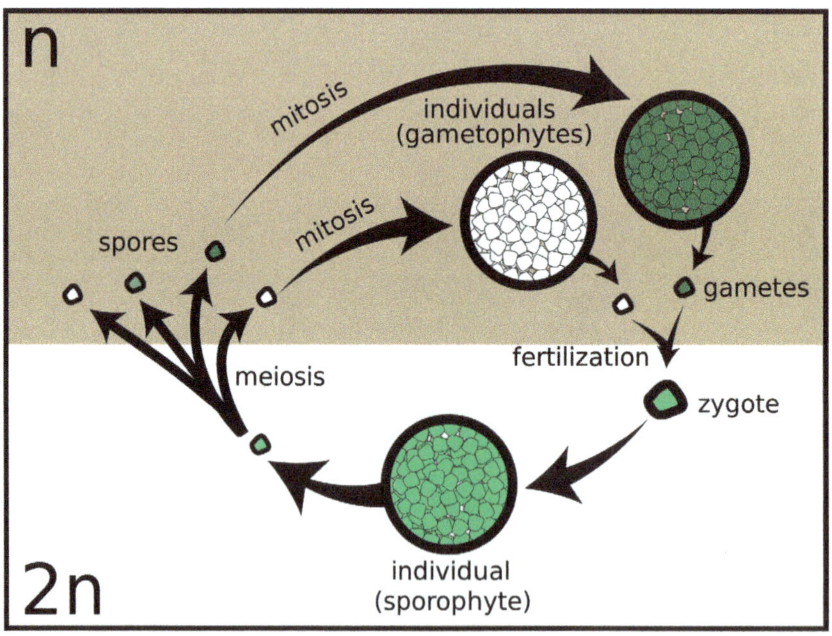

Sporic meiosis.

In **sporic meiosis** (also commonly known as **intermediary meiosis**), the zygote divides mitotically to produce a multicellular diploid sporophyte. The sporophyte creates spores via meiosis which *also* then divide mitotically producing haploid individuals called gametophytes. The gametophytes produce gametes via mitosis. In some plants the gametophyte is not only small-sized but also short-lived; in other plants and many algae, the gametophyte is the dominant stage of the life cycle.

Haplodiplonts are:

- In archaeplastidans: red algae (which have two sporophyte generations), some green algae (e.g., *Ulva*), land plants
- In stramenopiles: most brown algae
- In rhizarians: many foraminiferans, plasmodiophoromycetes
- In amoebozoa: myxogastrids
- In opisthokonts: some fungi (some chytrids, some ascomycetes like the brewer's yeast)
- Other eukaryotes: haptophytes

Some animals have a sex-determination system called haplodiploid, but this is not related to the haplodiplontic life cycle.

Vegetative Meiosis

Some red algae (such as *Bonnemaisonia* and *Lemanea*) and green algae (such as *Prasiola*) have **vegetative meiosis**, also called **somatic meiosis**, which is a rare phenomenon. Vegetative meiosis can occur in haplodiplontic and also in diplontic life cycles. The gametophytes remain attached to and part of the sporophyte. Vegetative (non-reproductive) diploid cells undergo meiosis, generating vegetative haploid cells. These undergo many mitosis, and produces gametes.

A different phenomenon, called **vegetative diploidization**, a type of apomixis, occurs in some brown algae (e.g., *Elachista stellaris*). Cells in a haploid part of the plant spontaneously duplicate their chromosomes to produce diploid tissue.

Evolution

The primitive type of life cycle probably had haploid individuals with asexual reproduction. Bacteria and archaea exhibit a life cycle like this, and some eukaryotes apparently do too (e.g., Cryptophyta, Choanoflagellata, many Euglenozoa, many Amoebozoa, some red algae, some green algae, the imperfect fungi, some rotifers and many other groups, not necessarily haploid). However, these eukaryotes probably are not primitively asexual, but have lost their sexual reproduction, or it just was not observed yet. Many eukaryotes (including animals and plants) exhibit asexual reproduction, which may be facultative or obligate in the life cycle, with sexual reproduction occurring more or less frequently.

ALTERNATION OF GENERATIONS

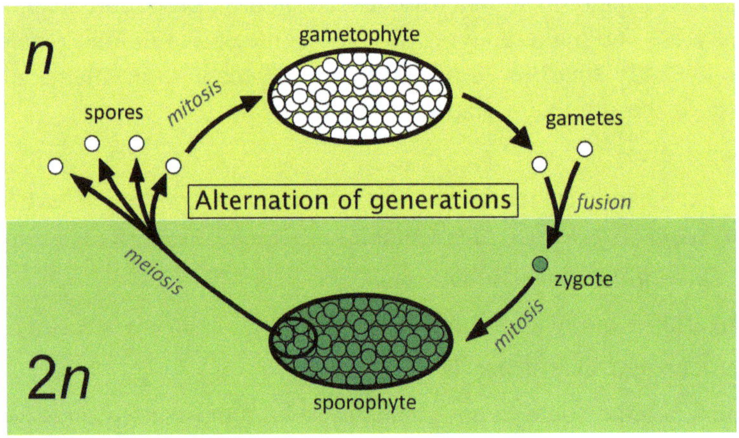

Diagram showing the alternation of generations between a diploid sporophyte (bottom) and a haploid gametophyte (top).

Alternation of generations (also known as **metagenesis**) is a term used to describe the type of life cycle that occurs in those plants and algae in the Archaeplastida and the Heterokontophyta that have distinct sexual haploid and asexual diploid stages. In these groups, a multicellular gametophyte, which is haploid with n chromosomes, alternates with a multicellular sporophyte, which is diploid with $2n$ chromosomes, made up of n pairs. A mature sporophyte produces spores by meiosis, a process which reduces the number of chromosomes to half, from $2n$ to n.

The haploid spores germinate and grow into a haploid gametophyte. At maturity, the gametophyte produces gametes by mitosis, which does not alter the number of chromosomes. Two gametes (originating from different organisms of the same species or from the same organism) fuse to produce a zygote, which develops into a diploid sporophyte. This cycle, from gametophyte to gametophyte (or equally from sporophyte to sporophyte), is the way in which all land plants and many algae undergo sexual reproduction.

The relationship between the sporophyte and gametophyte varies among different groups of plants. In those algae which have alternation of generations, the sporophyte and gametophyte are separate independent organisms, which may or may not have a similar appearance. In liverworts, mosses and hornworts, the sporophyte is less well developed than the gametophyte and is largely dependent on it. Although moss and hornwort sporophytes can photosynthesise, they require additional photosynthate from the gametophyte to sustain growth and spore development and depend on it for supply of water, mineral nutrients and nitrogen. By contrast, in all modern vascular plants the gametophyte is less well developed than the sporophyte, although their Devonian ancestors had gametophytes and sporophytes of approximately equivalent complexity. In ferns the gametophyte is a small flattened autotrophic prothallus on which the young sporophyte is briefly dependent for its nutrition. In flowering plants, the reduction of the gametophyte is much more extreme; it consists of just a few cells which grow entirely inside the sporophyte.

All animals develop differently. A mature animal is diploid and so is, in one sense, equivalent to a sporophyte. However, an animal *directly* produces haploid gametes by meiosis. No haploid spores capable of dividing are produced, so neither is a haploid gametophyte. There is no alternation between diploid and haploid forms.

Life cycles of plants and algae with alternating haploid and diploid multicellular stages are referred to as **diplohaplontic** (the equivalent terms **haplodiplontic**, **diplobiontic** or **dibiontic** are also in use). Life cycles, such as those of animals, in which there is only a diploid multicellular stage are referred to as **diplontic**. Life cycles in which there is only a haploid multicellular stage are referred to as **haplontic**.

Alternation of generations is defined as the alternation of multicellular diploid and haploid forms in the organism's life cycle, regardless of whether or not these forms are free-living. In some species, such as the alga *Ulva lactuca*, the diploid and haploid forms are indeed both free-living independent organisms, essentially identical in appearance and therefore said to be isomorphic. The free-swimming, haploid gametes form a diploid zygote which germinates into a multicellular diploid sporophyte. The sporophyte produces free-swimming haploid spores by meiosis that germinate into haploid gametophytes.

However, in some other groups, either the sporophyte or the gametophyte is very much reduced and is incapable of free living. For example, in all bryophytes the gametophyte generation is dominant and the sporophyte is dependent on it. By contrast, in all modern vascular land plants the gametophytes are strongly reduced, although the fossil evidence indicates that they were derived from isomorphic ancestors. In seed plants, the female gametophyte develops totally within the sporophyte which protects and nurtures it and the embryo sporophyte that it produces. The pollen grains, which are the male gametophytes, are reduced to only a few cells (just three cells in many cases). Here the notion of two generations is less obvious; as Bateman & Dimichele say [s]porophyte and gametophyte effectively function as a single organism. The alternative term 'alternation of phases' may then be more appropriate.

Debates about alternation of generations in the early twentieth century can be confusing because various ways of classifying generations co-exist (sexual vs. asexual, gametophyte vs. sporophyte, haploid vs. diploid, etc.).

Initially, Chamisso and Steenstrup described the succession of differently organized generations (sexual and asexual) in animals as "alternation of generations", while studying the development of tunicates, cnidarians and trematodes. This phenomenon is also known as heterogamy. Presently, the term alternation of generations is almost exclusively associated with the life cycles of plants, specifically with the alternation of haploid gametophytes and diploid sporophytes.

Wilhelm Hofmeister demonstrated the morphological alternation of generations in plants, between a spore-bearing generation (sporophyte) and a gamete-bearing generation (gametophyte). By that time, a debate emerged focusing on the origin of the asexual generation of land plants (i.e., the sporophyte) and is conventionally characterized as a conflict between theories of antithetic (Čelakovský, 1874) and homologous (Pringsheim, 1876) alternation of generations. Čelakovský coined the words sporophyte and gametophyte.

Eduard Strasburger (1874) discovered the alternation between diploid and haploid nuclear phases, also called cytological alternation of nuclear phases. Although most often coinciding, morphological alternation and nuclear phases alternation are sometimes independent of one another, e.g., in many red algae, the same nuclear phase may correspond to two diverse morphological generations. In some ferns which lost sexual reproduction, there is no change in nuclear phase, but the alternation of generations is maintained.

Alternation of Generations in Plants

Fundamental Elements

The diagram above shows the fundamental elements of the alternation of generations in plants. The many variations found in different groups of plants are described by use of these concepts later in the article. Starting from the right of the diagram, the processes involved are as follows:

- Two single-celled haploid gametes, each containing n unpaired chromosomes, fuse to form a single-celled diploid zygote, which now contains n pairs of chromosomes, i.e. $2n$ chromosomes in total.
- The single-celled diploid zygote germinates, dividing by the normal process (mitosis), which maintains the number of chromosomes at $2n$. The result is a multi-cellular diploid organism, called the *sporophyte* (because at maturity it produces spores).
- When it reaches maturity, the sporophyte produces one or more **sporangia** (singular: sporangium) which are the organs that produce diploid spore mother cells (sporocytes). These divide by a special process (meiosis) that reduces the number of chromosomes by a half. This initially results in four single-celled haploid spores, each containing n unpaired chromosomes.
- The single-celled haploid spore germinates, dividing by the normal process (mitosis), which maintains the number of chromosomes at n. The result is a multi-cellular haploid organism, called the *gametophyte* (because it produces gametes at maturity).
- When it reaches maturity, the gametophyte produces one or more **gametangia** (singular: gametangium) which are the organs that produce haploid gametes. At least one kind of gamete possesses some mechanism for reaching another gamete in order to fuse with it.

The 'alternation of generations' in the life cycle is thus between a diploid ($2n$) generation of sporophytes and a haploid (n) generation of gametophytes.

The situation is quite different from that in all animals, where the fundamental process is that a diploid ($2n$) individual *directly* produces haploid (n) gametes by meiosis. Spores (i.e. haploid cells which are able to undergo mitosis) are not produced, so neither is a haploid multi-cellular organism. The single-celled gametes are the only entities which are haploid.

Gametophyte of the fern *Onoclea sensibilis* (the flat thallus at the bottom of the picture) with a descendant sporophyte beginning to grow from it (the small frond at the top of the picture).

Variations

The diagram shown above is a good representation of the life cycle of some multi-cellular algae (e.g. the genus *Cladophora*) which have sporophytes and gametophytes of almost identical appearance and which do not have different kinds of spores or gametes.

However, there are many possible variations on the fundamental elements of a life cycle which has alternation of generations. Each variation may occur separately or in combination, resulting in a bewildering variety of life cycles. The terms used by botanists in describing these life cycles can be equally bewildering. As Bateman and Dimichele say "[...] the alternation of generations has become a terminological morass; often, one term represents several concepts or one concept is represented by several terms."

Possible variations are:

- Relative importance of the sporophyte and the gametophyte.
    - *Equal* (**homomorphy** or **isomorphy**). Filamentous algae of the genus *Cladophora*, which are predominantly found in fresh water, have diploid sporophytes and haploid gametophytes which are externally indistinguishable. No living land plant has equally dominant sporophytes and gametophytes, although some theories of the evolution of alternation of generations suggest that ancestral land plants did.
    - *Unequal* (**heteromorphy** or **anisomorphy**).

Gametophyte of *Mnium hornum*, a moss.

- *Dominant gametophyte* (**gametophytic**).
- In liverworts, mosses and hornworts, the dominant form is the haploid gametophyte. The diploid sporophyte is not capable of an independent existence, gaining most of its nutrition from the parent gametophyte, and having no chlorophyll when mature.

Sporophyte of *Blechnum discolor*, a fern.

- *Dominant sporophyte* (**sporophytic**).
- In ferns, both the sporophyte and the gametophyte are capable of living independently, but the dominant form is the diploid sporophyte. The haploid gametophyte is much smaller and simpler in structure. In seed plants, the gametophyte is even more reduced (at the minimum to only three cells), gaining all its nutrition from the sporophyte. The extreme reduction in the size of the gametophyte and its retention within the sporophyte means that when applied

to seed plants the term 'alternation of generations' is somewhat misleading: "[s] porophyte and gametophyte effectively function as a single organism". Some authors have preferred the term 'alternation of phases'.

- Differentiation of the gametes.
  - *Both gametes the same* (**isogamy**).
  - Like other species of *Cladophora*, *C. callicoma* has flagellated gametes which are identical in appearance and ability to move.
  - *Gametes of two distinct sizes* (**anisogamy**).
  - *Both of similar motility.*
  - Species of *Ulva*, the sea lettuce, have gametes which all have two flagella and so are motile. However they are of two sizes: larger 'female' gametes and smaller 'male' gametes.
  - *One large and sessile, one small and motile* (**oogamy**). The larger sessile megagametes are eggs (ova), and smaller motile microgametes are sperm (spermatozoa, spermatozoids). The degree of motility of the sperm may be very limited (as in the case of flowering plants) but all are able to move towards the sessile eggs. When (as is almost always the case) the sperm and eggs are produced in different kinds of gametangia, the sperm-producing ones are called **antheridia** (singular antheridium) and the egg-producing ones **archegonia** (singular archegonium).

Gametophyte of *Pellia epiphylla* with sporophytes growing from the remains of archegonia.

- *Antheridia and archegonia occur on the same gametophyte,* which is then called **monoicous**. (Many sources, including those concerned with bryophytes, use the term 'monoecious' for this situation and 'dioecious' for the opposite. Here 'monoecious' and 'dioecious' are used only for sporophytes.) The liverwort *Pellia epiphylla* has the gametophyte as the dominant generation. It is monoicous: the small reddish sperm-producing antheridia are scattered along the midrib while the egg-producing archegonia grow nearer the tips of divisions of the plant.

- *Antheridia and archegonia occur on different gametophytes,* which are then called **dioicous**. The moss *Mnium hornum* has the gametophyte as the dominant generation. It is dioicous: male plants produce only antheridia in terminal rosettes, female plants produce only archegonia in the form of stalked capsules. Seed plant gametophytes are also dioicous. However, the parent sporophyte may be monoecious, producing both male and female gametophytes or dioecious, producing gametophytes of one gender only. Seed plant gametophytes are extremely reduced in size; the archegonium consists only of a small number of cells, and the entire male gametophyte may be represented by only two cells.

■ Differentiation of the spores.

- *All spores the same size* (**homospory** or isospory). Horsetails (species of *Equisetum*) have spores which are all of the same size.
- *Spores of two distinct sizes* (**heterospory** or anisospory): larger **megaspores** and smaller **microspores**. When the two kinds of spore are produced in different kinds of sporangia, these are called **megasporangia** and **microsporangia**. A megaspore often (but not always) develops at the expense of the other three cells resulting from meiosis, which abort.
- *Megasporangia and microsporangia occur on the same sporophyte,* which is then called **monoecious**.
- Most flowering plants fall into this category. Thus the flower of a lily contains six stamens (the microsporangia) which produce microspores which develop into pollen grains (the microgametophytes), and three fused carpels which produce integumented megasporangia (ovules) each of which produces a megaspore which develops inside the megasporangium to produce the megagametophyte. In other plants, such as hazel, some flowers have only stamens, others only carpels, but the same plant (i.e. sporophyte) has both kinds of flower and so is monoecious.

Flowers of European Holly, a dioecious species: male above, female below (leaves cut to show flowers more clearly).

- *Megasporangia and microsporangia occur on different sporophytes,* which are then called **dioecious**.
- An individual tree of the European holly (*Ilex aquifolium*) produces either 'male' flowers which have only functional stamens (microsporangia) producing microspores which develop into pollen grains (microgametophytes) or 'female' flowers which have only functional carpels producing integumented megasporangia (ovules) that contain a megaspore that develops into a multicellular megagametophyte.

There are some correlations between these variations, but they are just that, correlations, and not absolute. For example, in flowering plants, microspores ultimately produce microgametes (sperm) and megaspores ultimately produce megagametes (eggs). However, in ferns and their allies there are groups with undifferentiated spores but differentiated gametophytes. For example, the fern *Ceratopteris thalictrioides* has spores of only one kind, which vary continuously in size. Smaller spores tend to germinate into gametophytes which produce only sperm-producing antheridia.

A Complex Life Cycle

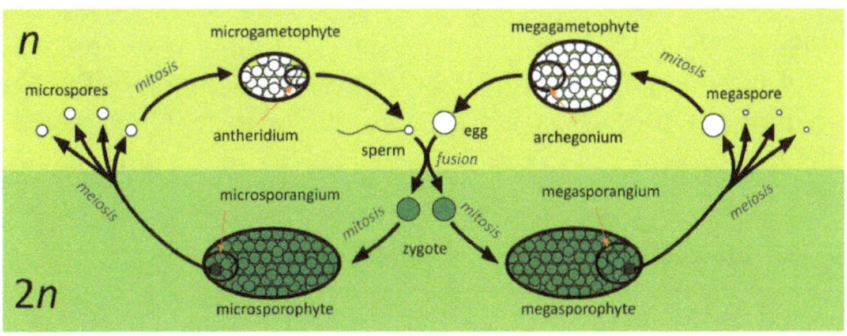

Graphic referred in text.

The diagram shows the alternation of generations in a species which is heteromorphic, sporophytic, oogametic, dioicous, heterosporic and dioecious. A seed plant example might be a willow tree (most species of the genus *Salix* are dioecious).Starting in the centre of the diagram, the processes involved are:

- An immobile egg, contained in the archegonium, fuses with a mobile sperm, released from an antheridium. The resulting zygote is either '**male'** or '**female'**.
    - A '**male'** zygote develops by mitosis into a microsporophyte, which at maturity produces one or more microsporangia. Microspores develop within the microsporangium by meiosis. In a willow (like all seed plants) the zygote first develops into an embryo microsporophyte within the ovule (a megasporangium enclosed in one or more protective layers of tissue known as integument). At maturity, these structures become the seed. Later the seed is shed, germinates and grows into a mature tree. A 'male'

willow tree (a microsporophyte) produces flowers with only stamens, the anthers of which are the microsporangia.

- Microspores germinate producing microgametophytes; at maturity one or more antheridia are produced. Sperm develop within the antheridia. In a willow, microspores are not liberated from the anther (the microsporangium), but develop into pollen grains (microgametophytes) within it. The whole pollen grain is moved (e.g. by an insect or by the wind) to an ovule (megagametophyte), where a sperm is produced which moves down a pollen tube to reach the egg.
- A '**female**' zygote develops by mitosis into a megasporophyte, which at maturity produces one or more megasporangia. Megaspores develop within the megasporangium; typically one of the four spores produced by meiosis gains bulk at the expense of the remaining three, which disappear. 'Female' willow trees (megasporophytes) produce flowers with only carpels (modified leaves that bear the megasporangia).
- Megaspores germinate producing megagametophytes; at maturity one or more archegonia are produced. Eggs develop within the archegonia. The carpels of a willow produce ovules, megasporangia enclosed in integuments. Within each ovule, a megaspore develops by mitosis into a megagametophyte. An archegonium develops within the megagametophyte and produces an egg. The whole of the gametophytic 'generation' remains within the protection of the sporophyte except for pollen grains (which have been reduced to just three cells contained within the microspore wall).

Life Cycles of Different Plant Groups

The term "plants" is taken here to mean the Archaeplastida, i.e. the glaucophytes, red and green algae and land plants.

Alternation of generations occurs in almost all multicellular red and green algae, both freshwater forms (such as *Cladophora*) and seaweeds (such as *Ulva*). In most, the generations are homomorphic (isomorphic) and free-living. Some species of red algae have a complex triphasic alternation of generations, in which there is a gametophyte phase and two distinct sporophyte phases. For further information, see Red algae: Reproduction.

Land plants all have heteromorphic (anisomorphic) alternation of generations, in which the sporophyte and gametophyte are distinctly different. All bryophytes, i.e. liverworts, mosses and hornworts, have the gametophyte generation as the most conspicuous. As an illustration, consider a monoicous moss. Antheridia and archegonia develop on the mature plant (the gametophyte). In the presence of water, the biflagellate sperm from the antheridia swim to the archegonia and fertilisation occurs, leading to the production of a diploid sporophyte. The sporophyte grows up from the archegonium. Its body comprises a long stalk topped by a capsule within which spore-producing cells undergo meiosis to form haploid spores. Most mosses rely on the wind to disperse these spores, although *Splachnum sphaericum* is entomophilous, recruiting insects to disperse its spores. For further information, see Liverwort: Life cycle, Moss: Life cycle, Hornwort: Life cycle.

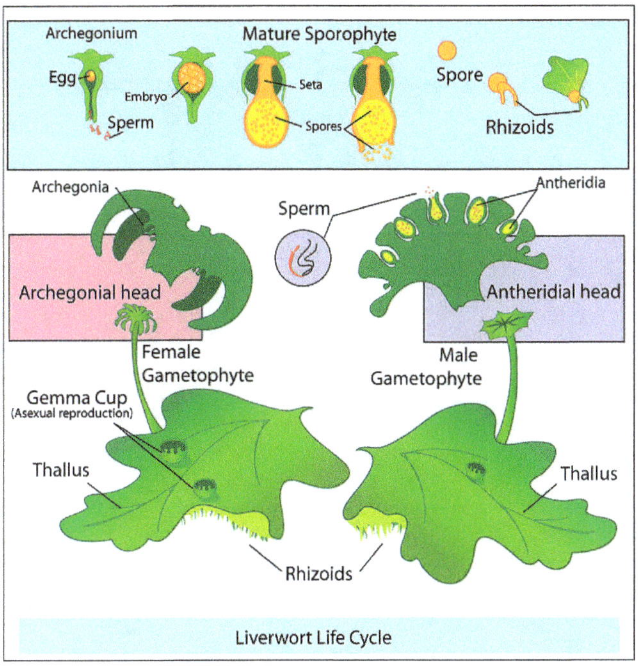

Diagram of alternation of generations in liverworts.

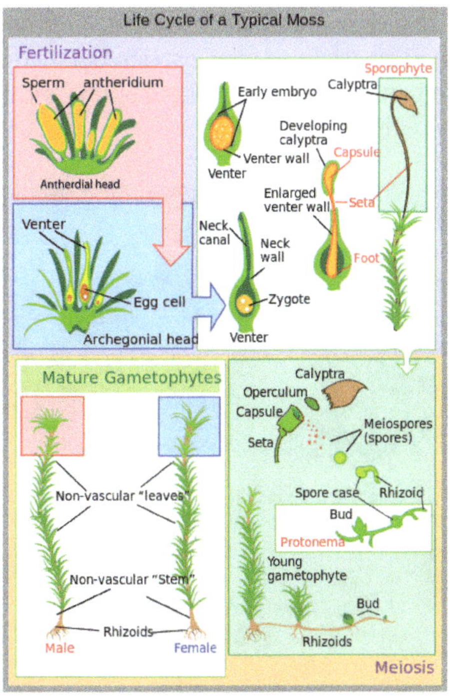

Moss life cycle diagram

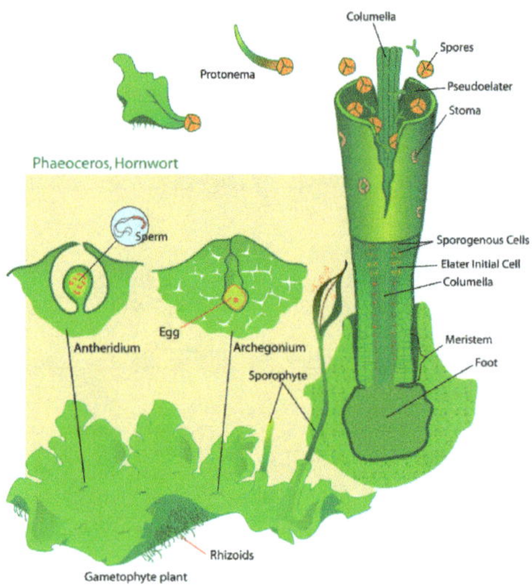

Hornwort life cycle diagram.

In ferns and their allies, including clubmosses and horsetails, the conspicuous plant observed in the field is the diploid sporophyte. The haploid spores develop in sori on the underside of the fronds and are dispersed by the wind (or in some cases, by floating on water). If conditions are right, a spore will germinate and grow into a rather inconspicuous plant body called a prothallus. The haploid prothallus does not resemble the sporophyte, and as such ferns and their allies have a heteromorphic alternation of generations. The prothallus is short-lived, but carries out sexual reproduction, producing the diploid zygote that then grows out of the prothallus as the sporophyte. For further information, see Fern: Life cycle.

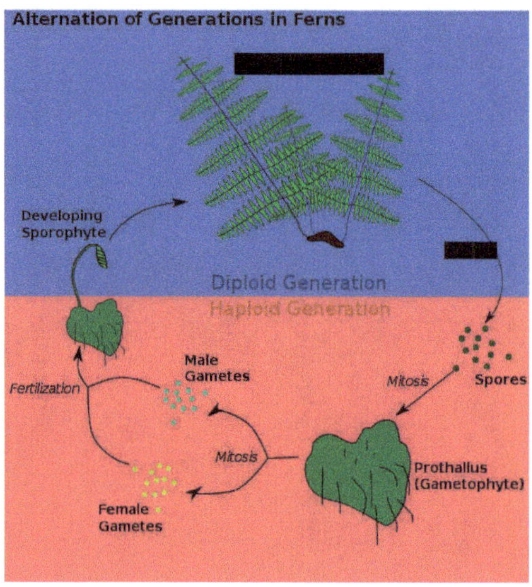

Diagram of alternation of generations in ferns.

A gametophyte (prothallus) of *Dicksonia sp.*

A sporophyte of *Dicksonia antarctica.*

In the spermatophytes, the seed plants, the sporophyte is the dominant multicellular phase; the gametophytes are strongly reduced in size and very different in morphology. The entire gametophyte generation, with the sole exception of pollen grains (microgametophytes), is contained within the sporophyte. The life cycle of a dioecious flowering plant (angiosperm), the willow, has been outlined in some detail in an earlier section (A complex life cycle). The life cycle of a gymnosperm is similar. However, flowering plants have in addition a phenomenon called 'double fertilization'. Two sperm nuclei from a pollen grain (the microgametophyte), rather than a single sperm, enter the archegonium of the megagametophyte; one fuses with the egg nucleus to form the zygote, the other fuses with two other nuclei of the gametophyte to form 'endosperm', which nourishes the developing embryo. For further information, see Double fertilization.

The underside of a *Dicksonia antarctica* frond showing the sori, or spore-producing structures.

Evolutionary Emergence of the Dominant Diploid Phase

It has been proposed that the basis for the emergence of the diploid phase of the life cycle (sporophyte) as the dominant phase (e.g. as in vascular plants) is that diploidy allows masking of the expression of deleterious mutations through genetic complementation. Thus if one of the parental genomes in the diploid cells contained mutations leading to defects in one or more gene products, these deficiencies could be compensated for by the other parental genome (which nevertheless may have its own defects in other genes). As the diploid phase was becoming predominant, the masking effect likely allowed genome size, and hence information content, to increase without the constraint of having to improve accuracy of DNA replication. The opportunity to increase information content at low cost was advantageous because it permitted new adaptations to be encoded. This view has been challenged, with evidence showing that selection is no more effective in the haploid than in the diploid phases of the lifecycle of mosses and angiosperms.

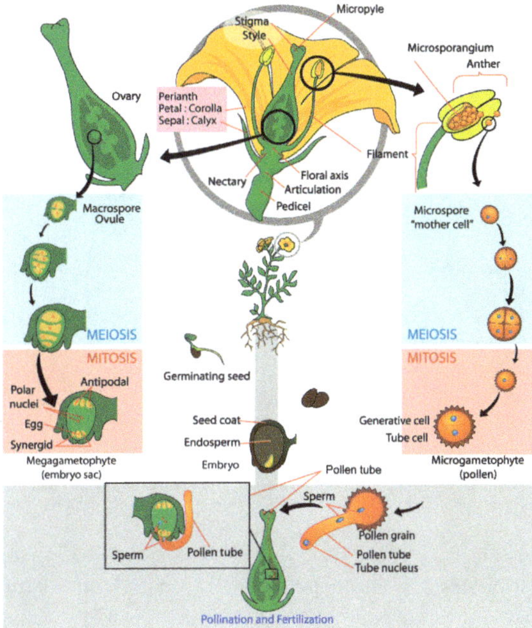

Angiosperm life cycle

Tip of tulip stamen showing pollen (microgametophytes)

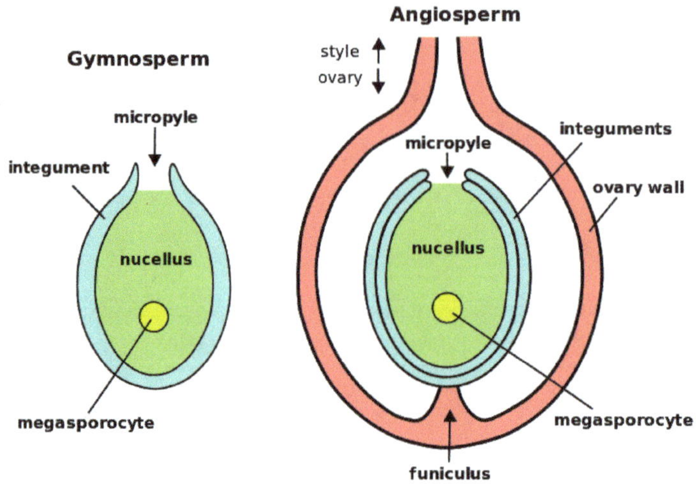

Plant ovules (megagametophytes): Gymnosperm ovule on left, angiosperm ovule (inside ovary) on right

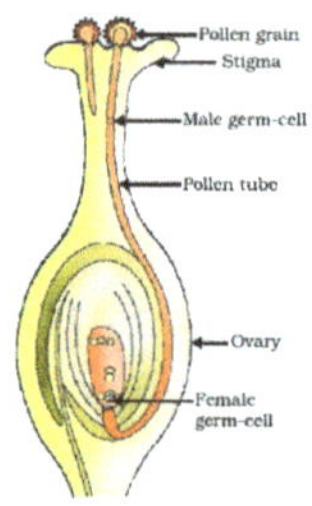

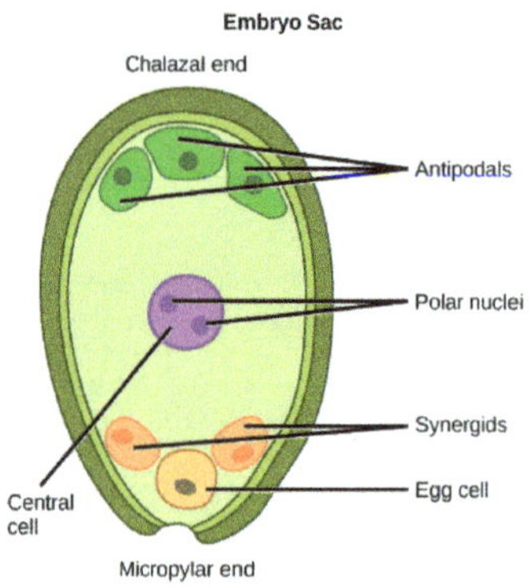

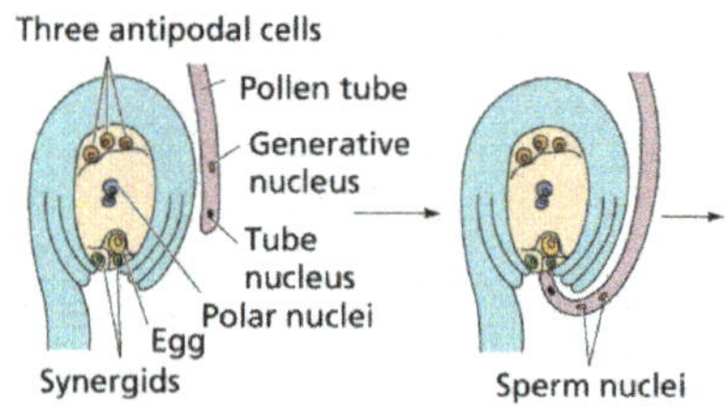

Double fertilization.

Similar Processes in Other Organisms

Chromalveolata

Some organisms currently classified in the Chromalveolata and thus not plants in the sense used here, exhibit alternation of generations. Foraminifera undergo a heteromorphic alternation of generations between haploid ***gamont*** and diploid ***agamont*** forms. The single-celled haploid organism is typically much larger than the diploid organism.

Fungi

Fungal mycelia are typically haploid. When mycelia of different mating types meet, they produce two multinucleate ball-shaped cells, which join via a "mating bridge". Nuclei move from one mycelium into the other, forming a ***heterokaryon*** (meaning "different nuclei"). This process is called ***plasmogamy***. Actual fusion to form diploid nuclei is called ***karyogamy***, and may not occur until sporangia are formed. Karogamy produces a diploid zygote, which is a short-lived sporophyte that soon undergoes meiosis to form haploid spores. When the spores germinate, they develop into new mycelia.

Slime moulds

The life cycle of slime moulds is very similar to that of fungi. Haploid spores germinate to form swarm cells or ***myxamoebae***. These fuse in a process referred to as *plasmogamy* and *karyogamy* to form a diploid zygote. The zygote develops into a plasmodium, and the mature plasmodium produces, depending on the species, one to many fruiting bodies containing haploid spores.

Animals

Alternation between a multicellular diploid and a multicellular haploid generation is never encountered in animals. In some animals, there is an alternation between parthenogenic and sexually reproductive phases (heterogamy). Both phases are diploid. This has sometimes been called alternation of generations, but is quite different.

REFERENCES

1. Thomas, R.J.; Stanton, D.S.; Longendorfer, D.H. & Farr, M.E. (1978), Physiological evaluation of the nutritional autonomy of a hornwort sporophyte, Botanical Gazette, 139 (3): 306–311, doi:10.1086/337006
2. Glime, J.M. (2007), Bryophyte Ecology: Vol. 1 Physiological Ecology (PDF), Michigan Technological University and the International Association of Bryologists, retrieved 2013-03-04
3. ^ Jump up to:[a] [b] Kerp, H.; Trewin, N.H. & Hass, H. (2003), New gametophytes from the Lower Devonian Rhynie Chert, Transactions of the Royal Society of Edinburgh: Earth Sciences, 94 (4): 411–428, doi:10.1017/S026359330000078X

4. Taylor, Kerp & Hass 2005
5. "Plant Science 4 U. Retrieved 5 July 2016.
6. Bateman & Dimichele 1994, p. 403
7. Stewart & Rothwell 1993
8. Haig, David (2008), "Homologous versus antithetic alternation of generations and the origin of sporophytes" (PDF), The Botanical Review, 74 (3): 395–418, doi:10.1007/s12229-008-9012-x, retrieved 2014-08-17
9. Svedelius, Nils (1927), Alternation of Generations in Relation to Reduction Division, Botanical Gazette, 83 (4): 362–384, doi:10.2307/2470766 (inactive 2017-01-15), JSTOR 2470766
10. Hofmeister, W. (1851), Vergleichende Untersuchungen der Keimung, Entfaltung und Fruchtbildildiung höherer Kryptogamen (Moose, Farne, Equisetaceen, Rhizocarpeen und Lycopodiaceen) und der Samenbildung der Coniferen (in German), Leipzig: F. Hofmeister, retrieved 2014-08-17. Translated as Currey, Frederick (1862), On the germination, development, and fructification of the higher Cryptogamia, and on the fructification of the Coniferæ, London: Robert Hardwicke, retrieved 2014-08-17
11. Feldmann, J. & Feldmann, G. (1942), Recherches sur les Bonnemaisoniacées et leur alternance de generations (PDF), Ann. Sci. Natl. Bot., ser. 11 (in French), 3: 75–175, p. 157
12. Feldmann, J. (1972), Les problèmes actuels de l'alternance de génerations chez les Algues, Bulletin de la Société Botanique de France (in French), 119: 7–38, doi:10.1080/00378941.1972.10839073
13. Schopfer, P.; Mohr, H. (1995). Physiology of Development. Plant physiology. Berlin: Springer. pp. 288–291. ISBN 3-540-58016-6.
14. Unless otherwise indicated, the material in the whole of this section is based on Foster & Gifford 1974, Sporne 1974a and Sporne 1974b.
15. Guiry & Guiry 2008
16. Bateman & Dimichele 1994, p. 347
17. Watson 1981, p. 2
18. Kirby 2001
19. Watson 1981, p. 33
20. Bell & Hemsley 2000, p. 104
21. Watson 1981, pp. 425–6
22. Watson 1981, pp. 287–8
23. Sporne 1974a, pp. 17–21.
24. Bateman & Dimichele 1994, pp. 350–1
25. Willows, Encyclopædia Britannica, XIX (11th ed.), New York: Encyclopædia Britannica, 1911, retrieved 2011-01-01

26. Bernstein, H.; Byers, G.S. & Michod, R.E. (1981), Evolution of sexual reproduction: Importance of DNA repair, complementation, and variation, The American Naturalist, 117 (4): 537–549, doi:10.1086/283734

27. Michod, R.E. & Gayley, T.W. (1992), Masking of mutations and the evolution of sex, The American Naturalist, 139 (4): 706–734, doi:10.1086/285354

28. Szövényi, Péter; Ricca, Mariana; Hock, Zsófia; Shaw, Jonathan A.; Shimizu, Kentaro K. & Wagner, Andreas (2013), Selection is no more efficient in haploid than in diploid life stages of an angiosperm and a moss, Molecular Biology and Evolution, 30 (8): 1929, doi:10.1093/molbev/mst095, PMID 23686659

29. Barnes et al. 2001, p. 321

30. Scott 1996, p. 35

CHAPTER 2

The Process of Meiosis

Chapter Contents

- Introduction
- Overview
- History
- Occurrence in Eukaryotic Life Cycles
- Process
- Phases
- Origin and Function
- Nondisjunction
- Meiosis in Plants and Animals
- Meiosis in Mammals
- Meiosis Vs. Mitosis
- G1/S Transition
- Biochemical Switches in the Cell Cycle
- Apoptosis

Meiosis is a process where a single cell divides twice to produce four cells containing half the original amount of genetic information. These cells are our sex cells – sperm in males, eggs in females.

INTRODUCTION

Meiosis is a specialized type of cell division that reduces the chromosome number by half, creating four haploid cells, each genetically distinct from the parent cell that gave rise to them. This process occurs in all sexually reproducing single-celled and multicellular eukaryotes, including animals, plants, and fungi. Errors in meiosis resulting in aneuploidy are the leading known cause of miscarriage and the most frequent genetic cause of developmental disabilities.

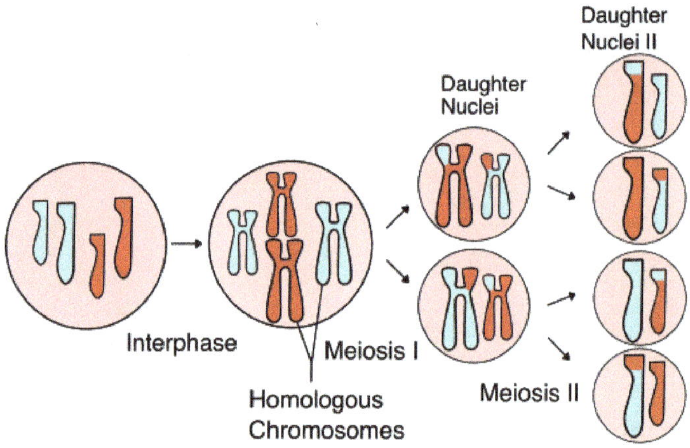

In meiosis, the chromosomes duplicate (during interphase) and homologous chromosomes exchange genetic information (chromosomal crossover) during the first division, called meiosis I. The daughter cells divide again in meiosis II, splitting up sister chromatids to form haploid gametes. Two gametes fuse during fertilization, creating a diploid cell with a complete set of paired chromosomes.

In meiosis, DNA replication is followed by two rounds of cell division to produce four potential daughter cells, each with half the number of chromosomes as the original parent cell. The two meiotic divisions are known as *Meiosis I* and *Meiosis II*. Before meiosis begins, during S phase of the cell cycle, the DNA of each chromosome is replicated so that it consists of two identical sister chromatids, which remain held together through sister chromatid cohesion. This S-phase can be referred to as premeiotic S-phase or meiotic S-phase. Immediately following DNA replication, meiotic cells enter a prolonged G2-like stage known as meiotic prophase. During this time, homologous chromosomes pair with each other and undergo genetic recombination, a programmed process in which DNA is cut and then repaired, which allows them to exchange some of their genetic information. A subset of recombination events results in crossovers, which create physical links known as chiasmata (singular: chiasma, for the Greek letter Chi (X)) between the homologous chromosomes. In most organisms, these links are essential to direct each pair of homologous chromosomes to segregate away from each other during Meiosis I, resulting in two haploid cells that have half the number of chromosomes as the parent cell. During Meiosis II, the cohesion between sister chromatids is released and they segregate from one another, as during mitosis. In some cases all four of the meiotic products form gametes such as sperm, spores, or pollen. In female animals, three of the four meiotic products are typically eliminated by extrusion into polar bodies, and only one cell develops to produce an ovum.

Because the number of chromosomes is halved during meiosis, gametes can fuse (i.e. fertilization) to form a diploid zygote that contains two copies of each chromosome, one from each parent. Thus, alternating cycles of meiosis and fertilization enable sexual reproduction, with successive generations maintaining the same number of chromosomes. For example, diploid human cells contain 23 pairs of chromosomes (46 total), half of maternal origin and half of paternal origin. Meiosis produces haploid gametes (ova or sperm) that contain one set of 23 chromosomes. When two gametes (an egg and a sperm) fuse, the resulting zygote is once again diploid, with the mother and father each contributing 23 chromosomes. This same pattern, but not the same number of chromosomes, occurs in all organisms that utilize meiosis.

OVERVIEW

While the process of meiosis is related to the more general cell division process of mitosis, it differs in two important respects:

recombination	meiosis	shuffles the genes between the two chromosomes in each pair (one received from each parent), producing recombinant chromosomes with unique genetic combinations in every gamete
	mitosis	occurs only if needed to repair DNA damage; usually occurs between identical sister chromatids and does not result in genetic changes
chromosome number (ploidy)	meiosis	produces four genetically unique cells, each with half the number of chromosomes as in the parent
	mitosis	produces two genetically identical cells, each with the same number of chromosomes as in the parent

Meiosis begins with a diploid cell, which contains two copies of each chromosome, termed homologs. First, the cell undergoes DNA replication, so each homolog now consists of two identical sister chromatids. Then each set of homologs pair with each other and exchange DNA by homologous recombination leading to physical connections (crossovers) between the homologs. In the first meiotic division, the homologs are segregated to separate daughter cells by the spindle apparatus. The cells then proceed to a second division without an intervening round of DNA replication. The sister chromatids are segregated to separate daughter cells to produce a total of four haploid cells. Female animals employ a slight variation on this pattern and produce one large ovum and two small polar bodies. Because of recombination, an individual chromatid can consist of a new combination of maternal and paternal DNA, resulting in offspring that are genetically distinct from either parent. Furthermore, an individual gamete can include an assortment of maternal, paternal, and recombinant chromatids. This genetic diversity resulting from sexual reproduction contributes to the variation in traits upon which natural selection can act.

Meiosis uses many of the same mechanisms as mitosis, the type of cell division used by eukaryotes to divide one cell into two identical daughter cells. In some plants, fungi, and protists meiosis results in the formation of spores: haploid cells that can divide vegetatively without undergoing fertilization. Some eukaryotes, like bdelloid rotifers, do not have the ability to carry out meiosis and have acquired the ability to reproduce by parthenogenesis.

Meiosis does not occur in archaea or bacteria, which generally reproduce via asexual processes such as binary fission. However, a sexual process known as horizontal gene transfer involves the transfer of DNA from one bacterium or archaeon to another and recombination of these DNA molecules of different parental origin.

HISTORY

Meiosis was discovered and described for the first time in sea urchin eggs in 1876 by the German biologist Oscar Hertwig. It was described again in 1883, at the level of chromosomes, by the Belgian zoologist Edouard Van Beneden, in *Ascaris* roundworm eggs. The significance of meiosis for reproduction and inheritance, however, was described only in 1890 by German biologist August Weismann, who noted that two cell divisions were necessary to transform one diploid cell into four haploid cells if the number of chromosomes had to be maintained. In 1911 the American geneticist Thomas Hunt Morgan detected crossovers in meiosis in the fruit fly *Drosophila melanogaster*, which helped to establish that genetic traits are transmitted on chromosomes.

The term meiosis (originally spelled "maiosis") was introduced to biology by J.B. Farmer and J.E.S. Moore in 1905:

We propose to apply the terms Maiosis or Maiotic phase to cover the whole series of nuclear changes included in the two divisions that were designated as Heterotype and Homotype by Flemming.

It is derived from the Greek word μείωσις, meaning lessening.

OCCURRENCE IN EUKARYOTIC LIFE CYCLES

Meiosis occurs in eukaryotic life cycles involving sexual reproduction, consisting of the constant cyclical process of meiosis and fertilization. This takes place alongside normal mitotic cell division. In multicellular organisms, there is an intermediary step between the diploid and haploid transition where the organism grows. At certain stages of the life cycle, germ cells produce gametes. Somatic cells make up the body of the organism and are not involved in gamete production.

Cycling meiosis and fertilization events produces a series of transitions back and forth between alternating haploid and diploid states. The organism phase of the life cycle can occur either during the diploid state (*gametic* or *diploid* life cycle), during the haploid state (*zygotic* or *haploid* life cycle), or both (*sporic* or *haplodiploid* life cycle, in which there are two distinct organism phases, one during the haploid state and the other during the diploid state). In this sense there are three types of life cycles that utilize sexual reproduction, differentiated by the location of the organism phase(s).

In the *gametic life cycle* or diplontic life cycle, of which humans are a part, the organism is diploid, grown from a diploid cell called the zygote. The organisms diploid germ-line stem cells undergo meiosis to create haploid gametes (the spermatozoa for males and ova for females), which fertilize to form the zygote. The diploid zygote undergoes repeated cellular division by mitosis to grow into the organism.

In the *zygotic life cycle* the organism is haploid instead, spawned by the proliferation and differentiation of a single haploid cell called the gamete. Two organisms of opposing sex contribute their haploid gametes to form a diploid zygote. The zygote undergoes meiosis immediately, creating four haploid cells. These cells undergo mitosis to create the organism. Many fungi and many protozoa utilize the zygotic life cycle.

Finally, in the *sporic life cycle*, the living organism alternates between haploid and diploid states. Consequently, this cycle is also known as the alternation of generations. The diploid organism's germ-line cells undergo meiosis to produce spores. The spores proliferate by mitosis, growing into a haploid organism. The haploid organism's gamete then combines with another haploid organism's gamete, creating the zygote. The zygote undergoes repeated mitosis and differentiation to become a diploid organism again. The sporic life cycle can be considered a fusion of the gametic and zygotic life cycles.

PROCESS

The preparatory steps that lead up to meiosis are identical in pattern and name to interphase of the mitotic cell cycle.

Interphase is divided into three phases:

- Growth 1 (G_1) phase: In this very active phase, the cell synthesizes its vast array of proteins, including the enzymes and structural proteins it will need for growth. In G1, each of the chromosomes consists of a single linear molecule of DNA.
- Synthesis (S) phase: The genetic material is replicated; each of the cell's chromosomes duplicates to become two identical sister chromatids attached at a centromere. This replication does not change the ploidy of the cell since the centromere number remains the same. The identical sister chromatids have not yet condensed into the densely packaged chromosomes visible with the light microscope. This will take place during prophase I in meiosis.
- Growth 2 (G2) phase: G2 phase as seen before mitosis is not present in meiosis. Meiotic prophase corresponds most closely to the G2 phase of the mitotic cell cycle.

Interphase is followed by meiosis I and then meiosis II. Meiosis I separates homologous chromosomes, each still made up of two sister chromatids, into two daughter cells, thus reducing the chromosome number by half. During meiosis II, sister chromatids decouple and the resultant daughter chromosomes are segregated into four daughter cells. For diploid organisms, the daughter cells resulting from meiosis are haploid and contain only one copy of each chromosome. In some species, cells enter a resting phase known as interkinesis between meiosis I and meiosis II.

Meiosis I and II are each divided into prophase, metaphase, anaphase, and telophase stages, similar in purpose to their analogous subphases in the mitotic cell cycle. Therefore, meiosis includes the stages of meiosis I (prophase I, metaphase I, anaphase I, telophase I) and meiosis II (prophase II, metaphase II, anaphase II, telophase II).

Meiosis generates gamete genetic diversity in two ways: (1) the independent orientation of homologous chromosome pairs along the metaphase plate during metaphase I and the subsequent separation of homologs during anaphase I allows a random and independent distribution of chromosomes to each daughter cell (and ultimately to gametes); and (2) physical exchange of homologous chromosomal regions by homologous recombination during prophase I results in new combinations of DNA within chromosomes.

During meiosis, specific genes are more highly transcribed. In addition to strong meiotic stage-specific expression of mRNA, there are also pervasive translational controls (e.g. selective usage of preformed mRNA), regulating the ultimate meiotic stage-specific protein expression of genes during meiosis. Thus, both transcriptional and translational controls determine the broad restructuring of meiotic cells needed to carry out meiosis.

PHASES

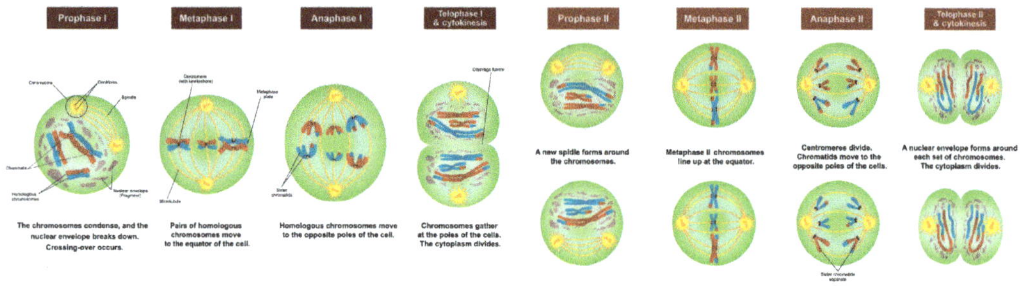

Diagram of the meiotic phases.

Meiosis is divided into meiosis I and meiosis II which are further divided into Karyokinesis I and Cytokinesis I & Karyokinesis II and Cytokinesis II respectively.

Meiosis I

Meiosis I segregates homologous chromosomes, which are joined as tetrads (2n, 4c), producing two haploid cells (n chromosomes, 23 in humans) which each contain chromatid pairs (1n, 2c). Because the ploidy is reduced from diploid to haploid, meiosis I is referred to as a *reductional division*. Meiosis II is an *equational division* analogous to mitosis, in which the sister chromatids are segregated, creating four haploid daughter cells (1n, 1c).

Prophase I

Prophase I is typically the longest phase of meiosis. During prophase I, homologous chromosomes pair and exchange DNA (homologous recombination). This often results in chromosomal crossover. This process is critical for pairing between homologous chromosomes and hence for accurate segregation of the chromosomes at the first meiosis division. The new combinations of DNA created during crossover are a significant source of genetic variation, and result in new combinations of alleles, which may be beneficial. The paired and replicated chromosomes are called bivalents or tetrads, which have two chromosomes and four chromatids, with one chromosome coming from each parent. The process of pairing the homologous chromosomes is called synapsis. At this stage, non-sister chromatids may cross-over at points called chiasmata (plural; singular chiasma).

Prophase I has historically been divided into a series of substages which are named according to the appearance of chromosomes.

Leptotene

The first stage of prophase I is the *leptotene* stage, also known as *leptonema*, from Greek words meaning "thin threads". In this stage of prophase I, individual chromosomes—each consisting of two sister chromatids—become "individualized" to form visible strands within the nucleus. The two sister chromatids closely associate and are visually indistinguishable from one another. During leptotene, lateral elements of the synaptonemal complex assemble. Leptotene is of very short duration and progressive condensation and coiling of chromosome fibers takes place.

Zygotene

The *zygotene* stage, also known as *zygonema*, from Greek words meaning "paired threads", occurs as the chromosomes approximately line up with each other into homologous chromosome pairs. In some organisms, this is called the bouquet stage because of the way the telomeres cluster at one end of the nucleus. At this stage, the synapsis (pairing/coming together) of homologous chromosomes takes place, facilitated by assembly of central element of the synaptonemal complex. Pairing is brought about in a zipper-like fashion and may start at the centromere (procentric), at the chromosome ends (proterminal), or at any other portion (intermediate). Individuals of a pair are equal in length and in position of the centromere. Thus pairing is highly specific and exact. The paired chromosomes are called bivalent or tetrad chromosomes.

Pachytene

The *pachytene* stage, also known as *pachynema*, from Greek words meaning "thick threads",. At this point a tetrad of the chromosomes has formed known as a bivalent. This is the stage when homologous recombination, including chromosomal crossover (crossing over), occurs. Nonsister chromatids of homologous chromosomes may exchange segments over regions of homology. Sex chromosomes, however, are not wholly identical, and only exchange information over a small region of homology. At the sites where exchange happens, chiasmata form. The exchange of information between the non-sister chromatids results in a recombination of information; each chromosome has the complete set of information it had before, and there are no gaps formed as a result of the process. Because the chromosomes cannot be distinguished in the synaptonemal complex, the actual act of crossing over is not perceivable through the microscope, and chiasmata are not visible until the next stage.

Diplotene

During the *diplotene* stage, also known as *diplonema*, from Greek words meaning "two threads", the synaptonemal complex degrades and homologous chromosomes separate from one another a little. The chromosomes themselves uncoil a bit, allowing some transcription of DNA. However, the homologous chromosomes of each bivalent remain tightly bound at chiasmata, the regions where crossing-over occurred. The chiasmata remain on the chromosomes until they are severed at the transition to anaphase I.

In mammalian and human fetal oogenesis all developing oocytes develop to this stage and are arrested before birth. This suspended state is referred to as the *dictyotene stage* or dictyate. It lasts until meiosis is resumed to prepare the oocyte for ovulation, which happens at puberty or even later.

Diakinesis

Chromosomes condense further during the *diakinesis* stage, from Greek words meaning "moving through". This is the first point in meiosis where the four parts of the tetrads are actually visible. Sites of crossing over entangle together, effectively overlapping, making chiasmata clearly visible. Other than this observation, the rest of the stage closely resembles prometaphase of mitosis; the nucleoli disappear, the nuclear membrane disintegrates into vesicles, and the meiotic spindle begins to form.

Synchronous Processes

During these stages, two centrosomes, containing a pair of centrioles in animal cells, migrate to the two poles of the cell. These centrosomes, which were duplicated during S-phase, function as microtubule organizing centers nucleating microtubules, which are essentially cellular ropes and poles. The microtubules invade the nuclear region after the nuclear envelope disintegrates, attaching to the chromosomes at the kinetochore. The kinetochore functions as a motor, pulling the chromosome along the attached microtubule toward the originating centrosome, like a train on a track. There are four kinetochores on each tetrad, but the pair of kinetochores on each sister chromatid fuses and functions as a unit during meiosis I.

Microtubules that attach to the kinetochores are known as *kinetochore microtubules*. Other microtubules will interact with microtubules from the opposite centrosome: these are called *nonkinetochore microtubules* or *polar microtubules*. A third type of microtubules, the aster microtubules, radiates from the centrosome into the cytoplasm or contacts components of the membrane skeleton.

Metaphase I

Homologous pairs move together along the metaphase plate: As *kinetochore microtubules* from both centrosomes attach to their respective kinetochores, the paired homologous chromosomes align along an equatorial plane that bisects the spindle, due to continuous counterbalancing forces exerted on the bivalents by the microtubules emanating from the two kinetochores of homologous chromosomes. The physical basis of the independent assortment of chromosomes is the random orientation of each bivalent along the metaphase plate, with respect to the orientation of the other bivalents along the same equatorial line. The protein complex cohesin holds sister chromatids together from the time of their replication until anaphase. In mitosis, the force of kinetochore microtubules pulling in opposite directions creates tension. The cell senses this tension and does not progress with anaphase until all the chromosomes are properly bi-oriented. In meiosis, establishing tension requires at least one crossover per chromosome pair in addition to cohesin between sister chromatids.

Anaphase I

Kinetochore microtubules shorten, pulling homologous chromosomes (which consist of a pair of sister chromatids) to opposite poles. Nonkinetochore microtubules lengthen, pushing the centrosomes farther apart. The cell elongates in preparation for division down the center.Unlike in mitosis, only the cohesin from the chromosome arms is degraded while the cohesin surrounding the centromere remains protected. This allows the sister chromatids to remain together while homologs are segregated.

Telophase I

The first meiotic division effectively ends when the chromosomes arrive at the poles. Each daughter cell now has half the number of chromosomes but each chromosome consists of a pair of chromatids. The microtubules that make up the spindle network disappear, and a new nuclear membrane surrounds each haploid set. The chromosomes uncoil back into chromatin. Cytokinesis, the pinching of the cell membrane in animal cells or the formation of the cell wall in plant cells, occurs, completing the creation of two daughter cells. Sister chromatids remain attached during telophase I.

Cells may enter a period of rest known as interkinesis or interphase II. No DNA replication occurs during this stage.

Meiosis II

Meiosis II is the second meiotic division, and usually involves equational segregation, or separation of sister chromatids. Mechanically, the process is similar to mitosis, though its genetic results are fundamentally different. The end result is production of four haploid cells (n chromosomes, 23 in humans) from the two haploid cells (with n chromosomes, each consisting of two sister chromatids) produced in meiosis I. The four main steps of Meiosis II are: Prophase II, Metaphase II, Anaphase II, and Telophase II.

In **prophase II** we see the disappearance of the nucleoli and the nuclear envelope again as well as the shortening and thickening of the chromatids. Centrosomes move to the Polar Regions and arrange spindle fibers for the second meiotic division.

In **metaphase II**, the centromeres contain two kinetochores that attach to spindle fibers from the centrosomes at opposite poles. The new equatorial metaphase plate is rotated by 90 degrees when compared to meiosis I, perpendicular to the previous plate.

This is followed by **anaphase II**, in which the remaining centromeric cohesin is cleaved allowing the sister chromatids to segregate. The sister chromatids by convention are now called sister chromosomes as they move toward opposing poles.

The process ends with **telophase II**, which is similar to telophase I, and is marked by decondensation and lengthening of the chromosomes and the disassembly of the spindle. Nuclear envelopes reform and cleavage or cell plate formation eventually produces a total of four daughter cells, each with a haploid set of chromosomes.

Meiosis is now complete and ends up with four new daughter cells.

ORIGIN AND FUNCTION

The **origin and function of meiosis** are fundamental to understanding the evolution of sexual reproduction in Eukaryotes. There is no current consensus among biologists on the questions of how sex in Eukaryotes arose in evolution, what basic function sexual reproduction serves, and why it is maintained, given the basic two-fold cost of sex. It is clear that it evolved over 1.2 billion years ago, and that almost all species which are descendents of the original sexually reproducing species are still sexual reproducers, including plants, fungi, and animals.

Meiosis is a key event of the sexual cycle in Eukaryotes. It is the stage of the life cycle when a cell gives rise to two haploid cells (gametes) each having half as many chromosomes. Two such haploid gametes, arising from different individual organisms, fuse by the process of fertilization, thus completing the sexual cycle.

Meiosis is ubiquitous among eukaryotes. It occurs in single-celled organisms such as yeast, as well as in multicellular organisms, such as humans. Eukaryotes arose from prokaryotes more than 2.2 billion years ago and the earliest eukaryotes were likely single-celled organisms. To understand sex in eukaryotes, it is necessary to understand (1) how meiosis arose in single celled eukaryotes, and (2) the function of meiosis.

Origin of Meiosis

There are two conflicting theories on how meiosis arose. One is that meiosis evolved from prokaryotic sex as eukaryotes evolved from prokaryotes. The other is that meiosis arose from mitosis.

From Prokaryotic Sex

In prokaryotic sex, DNA from one prokaryote is released into the surrounding medium, is then taken up by another prokaryote and its information integrated into the DNA of the recipient prokaryote. In modern prokaryotes the donor DNA is transferred by conjugation, transduction or transformation. Transformation is hypothesized to be the ancestral mechanism, in which DNA from one prokaryote is released into the surrounding medium and then taken up by another prokaryotic cell. One theory on how meiosis arose is that it evolved from transformation.According to this view, the evolutionary transition from prokaryotic sex to eukaryotic sex was continuous.

Transformation, like meiosis, is a complex process requiring the function of numerous gene products. A key similarity between prokaryotic sex and eukaryotic sex is that DNA originating from two different individuals (parents) join up so that homologous sequences are aligned with each other, and this is followed by exchange of genetic information (a process called genetic recombination). After the new recombinant chromosome is formed it is passed on to progeny.

When genetic recombination occurs between DNA molecules originating from different parents, the recombination process is catalyzed in prokaryotes and eukaryotes by enzymes that have similar functions and that are evolutionarily related. One of the most important enzymes catalyzing this process in bacteria is referred to as RecA, and this enzyme has two functionally similar counterparts that act in eukaryotic meiosis, RAD51 and DMC1.

The evolution of meiosis from transformation happened as eukaryotes evolved through endosymbiosis between an anaerobic host cell, most likely to be an archaeon and an internalized aerobic bacterium, most likely to be a descendant of α-proteobacterium, a likely ancestor of extant mitochondria. A genomic analysis on 630 orthologous groups hints a close evolutionary relationship between alpha-proteobacterial and eukaryotic proteins, suggesting there was indeed genetic transfer and, therefore, transformation between the two cells. Furthermore, in extant organisms, the RecA gene orthologs which are crucial for meiotic recombination have a high sequence similarity to that of the RecA gene in proteobacteria or cyanobacteria. This indicates a gene flow involving RecA gene or its orthologs from pre-mitochondrial bacteria to ancestral eukaryotes. Since the ancestors of eukaryotes were then homogenized to be able to undergo transformation, the genes needed for transformation, including RecA found in eubacteria and Rad51 family found in archaea, were likely descended to early eukaryotes as well and mutated into the ancestral meiotic genes.

Support for the theory that meiosis arose from prokaryotic transformation comes from the increasing evidence that early diverging lineages of eukaryotes have the core genes for meiosis. This implies that the precursor to meiosis was already present in the prokaryotic ancestor of eukaryotes. For instance the common intestinal parasite *Giardia intestinalis*, a simple eukaryotic protozoan was, until recently, thought to be descended from an early diverging eukaryotic lineage that lacked sex. However, it has since been shown that *G. intestinalis* contains within its genome a core set of genes that function in meiosis, including five genes that function only in meiosis. In addition, *G. intestinalis* was recently found to undergo a specialized, sex-like process involving meiosis gene homologs. This evidence, and other similar examples, suggest that a primitive form of meiosis, was present in the common ancestor of all eukaryotes, an ancestor that arose from antecedent prokaryote.

From Mitosis

Mitosis is the normal process in eukaryotes for cell division, duplicating chromosomes and segregating one of the two copies into each of the two daughter cells, and contrast with meiosis. The mitosis theory states that meiosis evolved from mitosis. On this theory, early eukaryotes evolved mitosis first, became established and only then did meiosis and sexual reproduction arise.

Support for the idea that meiosis arose from mitosis is the observation that some features, such as the meiotic spindles that draw chromosome sets into separate daughter cells upon cell division, and processes regulating cell division, employ the same, or similar, molecular machinery. Yet there is no compelling evidence for a period in the early evolution of eukaryotes during which meiosis and accompanying sexual capability did not yet exist.

In addition, as noted by Wilkins and Holliday, there are four novel steps needed in meiosis that are not present in mitosis. These are: (1) pairing of homologous chromosomes, (2) extensive recombination between homologs; (3) suppression of sister chromatid separation in the first meiotic division; and (4) avoiding chromosome replication during the second meiotic division. Although the introduction of these steps seems to be complicated, Wilkins and Holliday argue that only one new step, homolog synapsis, that was particularly initiated in the evolution of meiosis from mitosis. Meanwhile, two of the other novel features appear to be simply modified, and extensive recombination could have evolved later.

Coevolution with Mitosis

If meiosis arose from prokaryotic transformation, during the early evolution of eukaryotes, mitosis and meiosis could have evolved in parallel, with both processes using common molecular components, where mitosis evolved from the molecular machinery used by prokaryotes for DNA replication and segregation, and meiosis evolved from the prokaryotic sexual process of transformation, but meiosis also made use of the evolving molecular machinery for DNA replication and segregation.

Function

Stress Induced Sex

Abundant evidence indicates that facultative sexual eukaryotes tend to undergo sexual reproduction under stressful conditions. For instance, the budding yeast *Saccharomyces cerevisiae* (a single-celled fungus) reproduces mitotically (asexually) as diploid cells when nutrients are abundant, but switches to meiosis (sexual reproduction) under starvation conditions. The unicellular green alga, *Chlamydomonas reinhardtii* grows as vegetative cells in nutrient rich growth medium, but depletion of a source of nitrogen in the medium leads to gamete fusion, zygote formation and meiosis. The fission yeast *Schizosaccharomyces pombe*, treated with H2O2 to cause oxidative stress, substantially increases the proportion of cells which undergo meiosis. The simple multicellular eukaryote *Volvox carteri* undergoes sex in response to oxidative stress or stress from heat shock. These examples, and others, suggest that, in simple single-celled and multicellular eukaryotes, meiosis is an adaptation to respond to stress.

Prokaryotic sex also appears to be an adaptation to stress. For instance, transformation occurs near the end of logarithmic growth, when amino acids become limiting in *Bacillus subtilis*, or in *Haemophilus influenzae* when cells are grown to the end of logarithmic phase. In *Streptococcus mutans* and other streptococci, transformation is associated with high cell density and biofilm formation. In *Streptococcus pneumoniae*, transformation is induced by the DNA damaging agent mitomycin C. These, and other, examples indicate that prokaryotic sex, like meiosis in simple eukaryotes, is an adaptation to stressful conditions. This observation suggests that the natural selection pressures maintaining meiosis in eukaryotes are similar to the selective pressures maintaining prokaryotic sex. This similarity suggests continuity, rather than a gap, in the evolution of sex from prokaryotes to eukaryotes.

Stress is, however, a general concept. What is it specifically about stress that needs to be overcome by meiosis? And what is the specific benefit provided by meiosis that enhances survival under stressful conditions?

DNA Repair

In one theory, meiosis is primarily an adaptation for repairing DNA damage. Environmental stresses often lead to oxidative stress within the cell, which is well known to cause DNA damage through the production of reactive forms of oxygen, known as reactive oxygen species (ROS). DNA damages, if not repaired, can kill a cell by blocking DNA replication, or transcription of essential genes.

When only one strand of the DNA is damaged, the lost information (nucleotide sequence) can ordinarily be recovered by repair processes that remove the damaged sequence and fill the resulting gap by copying from the opposite intact strand of the double helix. However, ROS also cause a type of damage that is difficult to repair, referred to as double-strand damage. One common example of double-strand damage is the double-strand break. In this case, genetic information (nucleotide sequence) is lost from both strands in the damaged region, and proper information can only be obtained from another intact chromosome homologous to the damage chromosome. The process that the cell uses to accurately accomplish this type of repair is called recombinational repair.

Meiosis is distinct from mitosis in that a central feature of meiosis is the alignment of homologous chromosomes followed by recombination between them. The two chromosomes which pair are referred to as non-sister chromosomes, since they did not arise simply from the replication of a parental chromosome. Recombination between non-sister chromosomes at meiosis is known to be a recombinational repair process that can repair double-strand breaks and other types of double-strand damage. In contrast, recombination between sister chromosomes cannot repair double-strand damages arising prior to the replication which produced them. Thus on this view, the adaptive advantage of meiosis is that it facilitates recombinational repair of DNA damage that is otherwise difficult to repair, and that occurs as a result of stress, particularly oxidative stress. If left unrepaired, this damage would likely be lethal to gametes and inhibit production of viable progeny.

Even in multicellular eukaryotes, such as humans, oxidative stress is a problem for cell survival. In this case, oxidative stress is a byproduct of oxidative cellular respiration occurring during metabolism in all cells. In humans, on average, about 50 DNA double-strand breaks occur per cell in each cell generation. Meiosis, which facilitates recombinational repair between non-sister chromosomes, can efficiently repair these prevalent damages in the DNA passed on to germ cells, and consequently prevent loss of fertility in humans. Thus with the theory that meiosis arose from prokaryotic sex, recombinational repair is the selective advantage of meiosis in both single celled eukaryotes and muticellular eukaryotes, such as humans.

An argument against this hypothesis is that adequate repair mechanisms including those involving recombination already exist in prokaryotes.Prokaryotes do have DNA repair mechanism enriched with recombinational repair, and the existence of prokaryotic life in severe environment indicates the extreme efficiency of this mechanism to help them survive many DNA damages related to the environment. This implies that an extra costly repair in the form of meiosis would be unnecessary. However, most of these mechanisms cannot be as accurate as and is possibly more mutagenic than the repair mechanism provided by meiosis. They primarily do not require a second homologous chromosome for the recombination that promotes a more extensive repair. Thus, despite the efficiency of recombinational repair involving sister chromatids, the repair still needs to be improved, and another type of repair is required. Moreover, due to the more extensive homologous recombinational repair in meiosis in comparison to the repair in mitosis, meiosis as a repair mechanism can accurately remove any damage that arises at any stage of the cell cycle more than mitotic repair mechanism can do and was, therefore, naturally selected. In contrast, the sister chromatid in mitotic recombination could have been exposed to similar amount of stress, and, thus, this type of recombination, instead of eliminating the damage, could actually spread the damage and decrease fitness.

Genetic Diversity

On the other view, stress is a signal to the cell that the environment is becoming adverse. Under this new condition, it may be beneficial to produce progeny that differ from the parent in their genetic make up. Among these varied progeny, some may be more adapted to the changed condition than their parents. Meiosis generates genetic variation in the diploid cell, in part by the exchange of genetic information between the pairs of chromosomes after they align (recombination). Thus, on this view, the advantage of meiosis is that it facilitates the generation of genomic diversity among progeny, allowing adaptation to adverse changes in the environment.

However, in the presence of a fairly stable environment, individuals surviving to reproductive age have genomes that function well in their current environment. They raise the question of why such individuals should risk shuffling their genes with those of another individual, as occurs during meiotic recombination? Considerations such as this have led many investigators to question whether genetic diversity is the adaptive advantage of sex.

NONDISJUNCTION

Nondisjunction is the failure of homologous chromosomes or sister chromatids to separate properly during cell division. There are three forms of nondisjunction: failure of a pair of homologous chromosomes to separate in meiosis I, failure of sister chromatids to separate during meiosis II, and failure of sister chromatids to separate during mitosis. Nondisjunction results in daughter cells with abnormal chromosome numbers (aneuploidy).

Calvin Bridges and Thomas Hunt Morgan are credited with discovering nondisjunction in *Drosophila melanogaster* sex chromosomes in the spring of 1910, while working in the Zoological Laboratory of Columbia University.

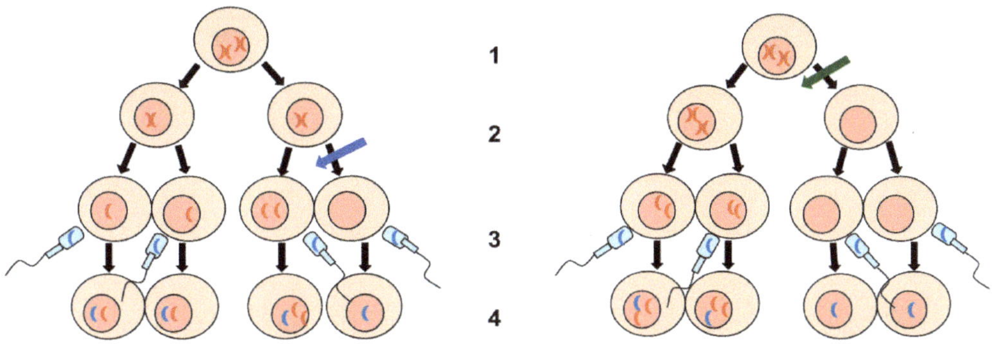

1. Meiosis I 2. Meiosis II 3. Fertilization 4. Zygote The left image at the blue arrow is nondisjunction taking place during meiosis II. The right image at the green arrow is nondisjunction taking place during meiosis I. Nondisjunction is when chromosomes fail to separate normally resulting in a gain or loss of chromosomes.

Types

In general, nondisjunction can occur in any form of cell division that involves ordered distribution of chromosomal material. Higher animals have three distinct forms of such cell divisions: Meiosis I and meiosis II are specialized forms of cell division occurring during generation of gametes (eggs and sperm) for sexual reproduction, mitosis is the form of cell division used by all other cells of the body.

Meiosis II

Ovulated eggs become arrested in metaphase II until fertilization triggers the second meiotic division. Similar to the segregation events of mitosis, the pairs of sister chromatids resulting from the separation of bivalents in meiosis I are further separated in anaphase of meiosis II. In oocytes, one sister chromatid is segregated into the second polar body, while the other stays inside the egg. During spermatogenesis, each meiotic division is symmetric such that each primary spermatocyte gives rise to 2 secondary spermatocytes after meiosis I, and eventually 4 spermatids after meiosis II. Meiosis II nondisjunction may also result in aneuploidy syndromes, but only to a much smaller extent than do segregation failures in meiosis I.

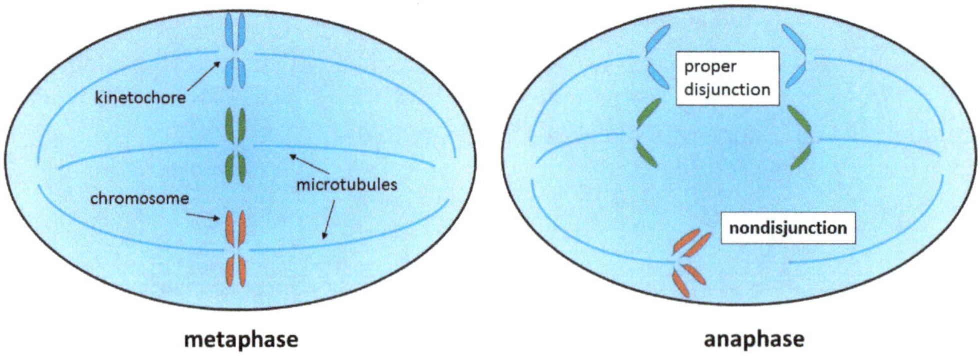

Nondisjunction of sister chromatids during mitosis:

Left: Metaphase of mitosis. Chromosome line up in the middle plane, the mitotic spindle forms and the kinetochores of sister chromatids attach to the microtubules.

Right: Anaphase of mitosis, where sister chromatids separate and the microtubules pull them in opposite directions.

The chromosome shown in *red* fails to separate properly, its sister chromatids stick together and get pulled to the same side, resulting in mitotic nondisjunction of this chromosome.

Mitosis

Division of somatic cells through mitosis is preceded by replication of the genetic material in S phase. As a result, each chromosome consists of two sister chromatids held together at the centromere. In the anaphase of mitosis, sister chromatids separate and migrate to opposite cell poles before the cell divides. Nondisjunction during mitosis leads to one daughter receiving both sister chromatids of the affected chromosome while the other gets none. This is known as a chromatin bridge or an anaphase bridge. Mitotic nondisjunction results in somatic mosaicism, since only daughter cells originating from the cell where the nondisjunction event has occurred will have an abnormal number of chromosomes. Nondisjunction during mitosis can contribute to the development of some forms of cancer, e.g. retinoblastoma (see below). Chromosome nondisjunction in mitosis can be atrributed to the inactivation of topoisomerase II, condensin, or separase. Meiotic nondisjunction has been well studied in *Saccharomyces cerevisiae*. This yeast undergoes mitosis similarly to other eukaryotes. Chromosome bridges occur when sister chromatids are held together post replication by DNA-DNA topological entanglement and the cohesion complex. During anaphase, cohesion is cleaved by separase. Topoisomerase II and condensin are responsible for removing catenations.

Molecular Mechanisms

Central Role of the Spindle Assembly Checkpoint

The spindle assembly checkpoint (SAC) is a molecular safe-guarding mechanism that governs proper chromosome segregation in eukaryotic cells. SAC inhibits progression into anaphase until all homologous chromosomes (bivalents) are properly aligned to the spindle apparatus. Only then, SAC releases its inhibition of the anaphase promoting complex (APC), which in turn irreversibly triggers progression through anaphase.

Sex-specific Differences in Meiosis

Surveys of cases of human aneuploidy syndromes have shown that most of them are maternally derived. This raises the question: Why is female meiosis more error prone? The most obvious difference between female oogenesis and male spermatogenesis is the prolonged arrest of oocytes in late stages of prophase I for many years up to several decades. Male gametes on the other hand quickly go through all stages of meiosis I and II. Another important difference between male and female meiosis concerns the frequency of recombination between homologous chromosomes: In the male, almost all chromosome pairs are joined by at least one crossover, while more than 10% of human oocytes contain at least one bivalent without any crossover event. Failures of recombination or inappropriately located crossovers have been well documented as contributors to the occurrence of nondisjunction in humans.

Age-related loss of Cohesin Ties

Due to the prolonged arrest of human oocytes, weakening of cohesive ties holding together chromosomes and reduced activity of the SAC may contribute to maternal age-related errors in segregation control. The cohesin complex is responsible for keeping together sister chromatids and provides binding sites for spindle attachment. Cohesin is loaded onto newly replicated chromosomes in oogonia during fetal development. Mature oocytes have only limited capacity for reloading cohesin after completion of S phase. The prolonged arrest of human oocytes prior to completion of meiosis I may therefore result in considerable loss of cohesin over time. Loss of cohesin is assumed to contribute to incorrect microtubule-kinetochore attachment and chromosome segregation errors during meiotic divisions.

Consequences

The result of this error is a cell with an imbalance of chromosomes. Such a cell is said to be aneuploid. Loss of a single chromosome (2n-1), in which the daughter cell(s) with the defect will have one chromosome missing from one of its pairs, is referred to as a monosomy. Gaining a single chromosome, in which the daughter cell(s) with the defect will have one chromosome in addition to its pairs is referred to as a trisomy. In the event that an aneuploidic gamete is fertilized, a number of syndromes might result.

Monosomy

The only known survivable monosomy in humans is Turner syndrome, where the affected individual is monosomic for the X chromosome (see below). Other monosomies are usually lethal during early fetal development, and survival is only possible if not all the cells of the body are affected in case of a mosaicism (see below), or if the normal number of chromosomes is restored via duplication of the single monosomic chromosome ("chromosome rescue").

Turner Syndrome (X Monosomy) (45, X0)

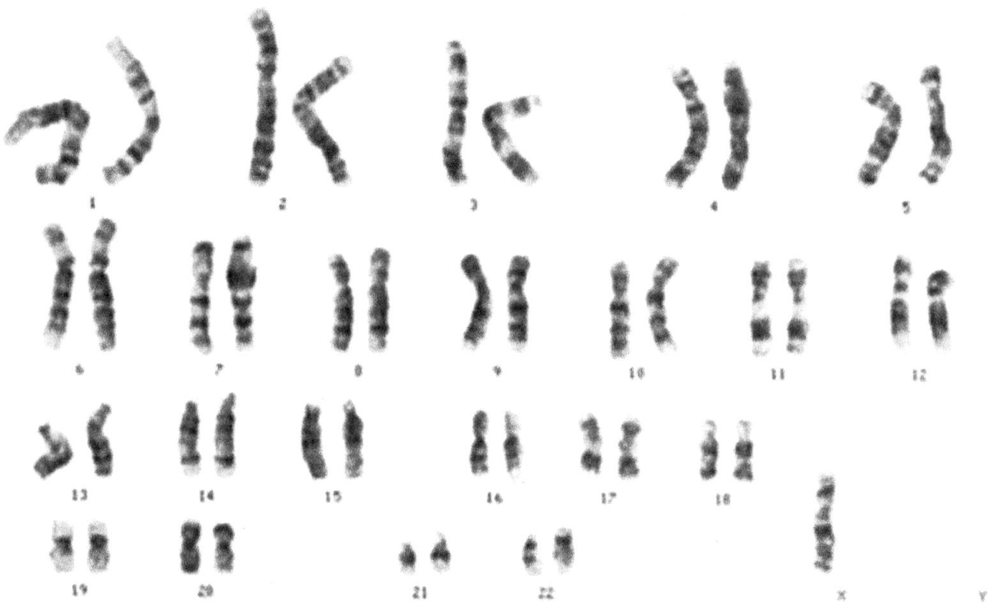

Karyotype of X monosomy (Turner syndrome)

This condition is characterized by the presence of **only one X chromosome** and no Y chromosome (see bottom right corner).

Complete loss of an entire X chromosome accounts for about half the cases of Turner syndrome. The importance of both X chromosomes during embryonic development is underscored by the observation that the overwhelming majority (>99%) of fetuses with only one X chromosome (karyotype 45, X0) are spontaneously aborted.

Autosomal Trisomy

The term autosomal trisomy means that a chromosome other than the sex chromosomes X and Y is present in 3 copies instead of the normal number of 2 in diploid cells.

Down Syndrome (Trisomy 21)

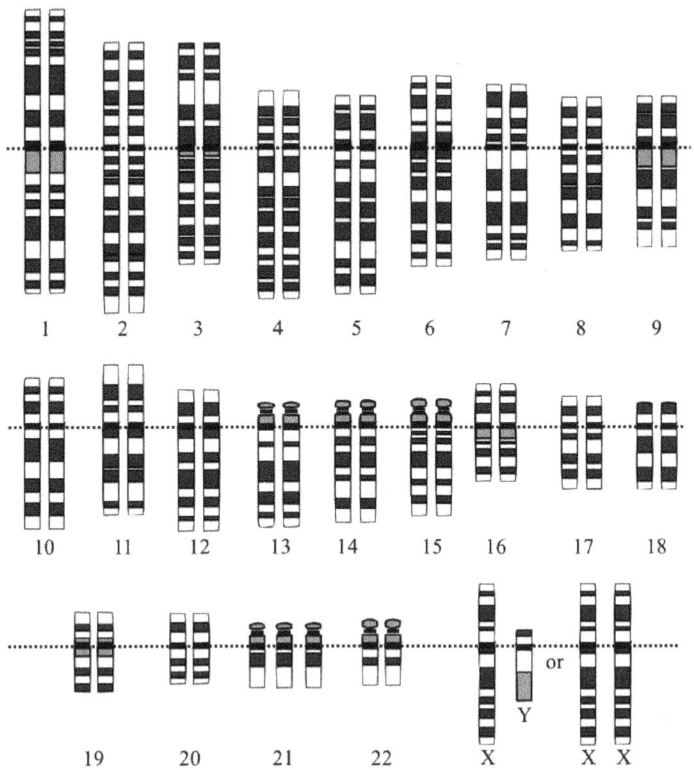

Karyotype of trisomy 21 (Down syndrome)

Note that chromosome 21 is present in 3 copies, while all other chromosomes show the normal diploid state with 2 copies. Most cases of trisomy of chromosome 21 are caused by a nondisjunction event during meiosis I (see text).

Down syndrome, a trisomy of chromosome 21, is the most common anomaly of chromosome number in humans. The majority of cases results from nondisjunction during maternal meiosis I. Trisomy occurs in at least 0.3% of newborns and in nearly 25% of spontaneous abortions. It is the leading cause of pregnancy wastage and is the most common known cause of mental retardation. It is well documented that advanced maternal age is associated with greater risk of meiotic nondisjunction leading to Down syndrome. This may be associated with the prolonged meiotic arrest of human oocytes potentially lasting for more than four decades.

Edwards Syndrome (Trisomy 18) and Patau Syndrome (Trisomy 13)

Human trisomies compatible with life birth, other than Down syndrome (trisomy 21), are Edwards syndrome (trisomy 18), and Patau syndrome (trisomy 13). Complete trisomies of other chromosomes are usually not viable and represent a relatively frequent cause of miscarriage. Only in rare cases of a mosaicism, the presence of a normal cell line, in addition to the trisomic cell line, may support the development of a viable trisomy of the other chromosomes.

Sex Chromosome Aneuploidy

The term *sex chromosome aneuploidy* summarizes conditions with an abnormal number of sex chromosomes, i.e. other than XX (female) or XY (male). Formally, X chromosome monosomy (Turner syndrome, see above) can also be classified as a form of sex chromosome aneuploidy.

Klinefelter Syndrome (47, XXY)

Klinefelter syndrome is the most common sex chromosome aneuploidy in humans. It represents the most frequent cause of hypogonadism and infertility in men. Most cases are caused by nondisjunction errors in paternal meiosis I.About eighty percent of individuals with this syndrome have one extra X chromosome resulting in the karyotype XXY. The remaining cases have either multiple additional sex chromosomes (48,XXXY; 48,XXYY; 49,XXXXY), mosaicism (46,XY/47,XXY), or structural chromosome abnormalities.

XYY Male (47, XYY)

The incidence of XYY syndrome is approximately 1 in 800-1000 male births. Many cases remain undiagnosed because of their normal appearance and fertility, and the absence of severe symptoms. The extra Y chromosome is usually a result of nondisjunction during paternal meiosis II.

Trisomy X (47,XXX)

Trisomy X is a form of sex chromosome aneuploidy where females have three instead of two X chromosomes. Most patients are only mildly affected by neuropsychological and physical symptoms. Studies examining the origin of the extra X chromosome observed that about 58-63% of cases were caused by nondisjunction in maternal meiosis I, 16-18% by nondisjunction in maternal meiosis II, and the remaining cases by post-zygotic, i.e. mitotic, nondisjunction.

Uniparental Disomy

Uniparental disomy denotes the situation where both chromosomes of a chromosome pair are inherited from the same parent and are therefore identical. This phenomenon most likely is the result of a pregnancy that started as a trisomy due to nondisjunction. Since most trisomies are lethal, the fetus only survives because it loses one of the three chromosomes and becomes disomic. Uniparental disomy of chromosome 15 is, for example, seen in some cases of Prader-Willi syndrome and Angelman syndrome.

Mosaicism Syndromes

Mosaicism syndromes can be caused by mitotic nondisjunction in early fetal development. As a consequence, the organism evolves as a mixture of cell lines with differing ploidy (number of chromosomes). Mosaicism may be present in some tissues, but not in others. Affected individuals may have a patchy or assymmetric appearance. Examples of mosaicism syndromes include Pallister-Killian syndrome and Hypomelanosis of Ito.

Mosaicism in Malignant Transformation

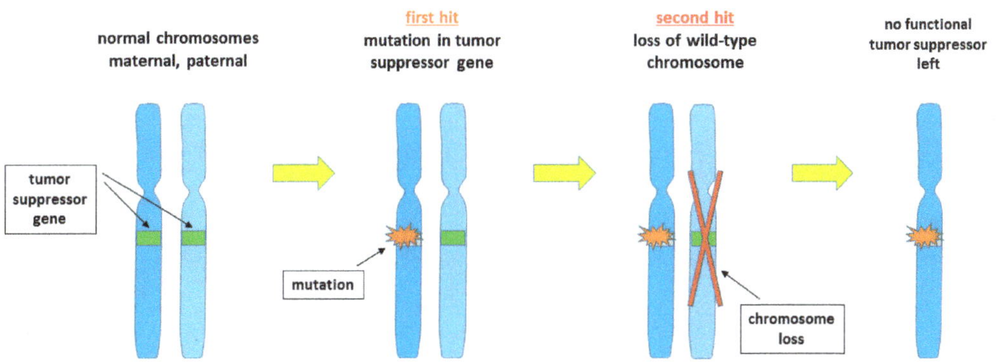

Loss of a tumor suppressor gene locus according to the two-hit model:

In the first hit, the tumor suppressor gene on one of the two chromosomes is affected by a mutation that makes the gene product non-functional. This mutation may arise spontaneously as a DNA replication error or may be induced by a DNA damaging agent. The second hit removes the remaining wild-type chromosome, for example through a **mitotic nondisjunction** event. There are several other potential mechanisms for each of the two steps, for example an additional mutation, an unbalanced translocation, or a gene deletion by recombination. As a result of the double lesion, the cell may become malignant because it is no longer able to express the tumor suppressor protein.

Development of cancer often involves multiple alterations of the cellular genome (Knudson hypothesis). Human retinoblastoma is a well studied example of a cancer type where mitotic nondisjunction can contribute to malignant transformation: Mutations of the RB1 gene, which is located on chromosome 13 and encodes the tumor suppressor retinoblastoma protein, can be detected by cytogenetic analysis in many cases of retinoblastoma. Mutations of the RB1 locus in one copy of chromosome 13 are sometimes accompanied by loss of the other wild-type chromosome 13 through mitotic nondisjunction. By this combination of lesions, affected cells completely lose expression of functioning tumor suppressor protein.

Diagnosis

Preimplantation Genetic Diagnosis

Pre-implantation genetic diagnosis (PGD or PIGD) is a technique used to identify genetically normal embryos and is useful for couples who have a family history of genetic disorders. This is an option for people choosing to procreate through IVF. PGD is considered difficult due to it being both time consuming and having success rates only comparable to routine IVF.

Karyotyping

Karyotyping involves performing an amniocentesis in order to study the cells of an unborn fetus. Light microscopy can be used to visually determine if aneuploidy is an issue.

Polar Body Diagnosis

Polar body diagnosis (PBD) can be use to detect maternally derived chromosomal aneuploidies as well as translocations in oocytes. The advantage of PBD over PGD is that it can be accomplished in a short amount of time. This is accomplished through zona drilling or laser drilling.

Blastomere Biopsy

Blastomere biopsy is a technique in which blastomeres are removed from the zona pellucida. It is commonly used to detect aneuploidy. Genetic analysis is conducted once the procedure is complete. Additional studies are needed to assess the risk associated with the procedure.

Lifestyle/Environmental Hazards

Exposure of spermatozoa to lifestyle, environmental and/or occupational hazards may increase the risk of aneuploidy. Cigarette smoke is a known aneugen (aneuploidy inducing agent). It is associated with increases in aneuploidy ranging from 1.5 to 3.0-fold. Other studies indicate factors such as alcohol consumption, occupational exposure to benzene, and exposure to the insecticides fenvalerate and carbaryl also increase aneuploidy.

MEIOSIS IN PLANTS AND ANIMALS

Meiosis occurs in all animals and plants. The end result, the production of gametes with half the number of chromosomes as the parent cell, is the same, but the detailed process is different. In animals, meiosis produces gametes directly. In land plants and some algae, there is an alternation of generations such that meiosis in the diploid sporophyte generation produces haploid spores. These spores multiply by mitosis, developing into the haploid gametophyte generation, which then gives rise to gametes directly (i.e. without further meiosis). In both animals and plants, the final stage is for the gametes to fuse, restoring the original number of chromosomes.

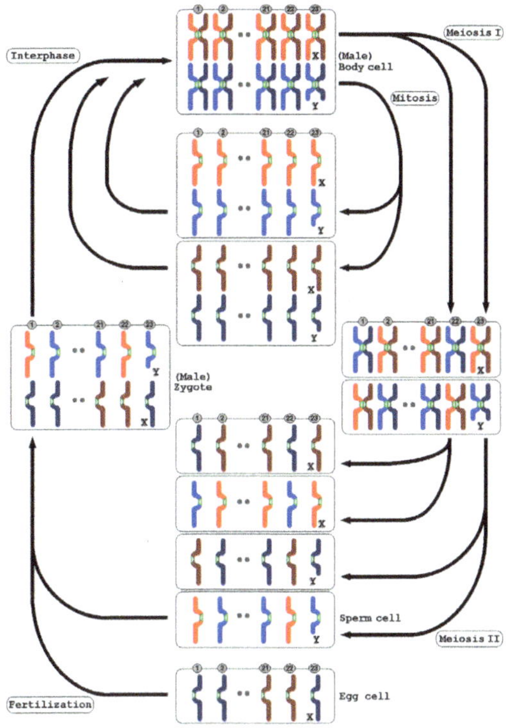

Overview of chromatides' and chromosomes' distribution within the mitotic and meiotic cycle of a male human cell

MEIOSIS IN MAMMALS

In females, meiosis occurs in cells known as oocytes (singular: oocyte). Each oocyte that initiates meiosis divides twice, unequally in each case. The first division produces a daughter cell that will undergo a second division, and a much smaller "polar body" that is extruded from the surface of the cell and does not divide further. Following Meiosis II, a "second polar body" is extruded, and the single remaining haploid cell enlarges to become an ovum. Since the first polar body normally disintegrates rather than dividing again, meiosis in female mammals results in three products, the oocyte and two polar bodies. However, before these divisions occur, these cells stop at the diplotene stage of meiosis I and lie dormant within a protective shell of somatic cells called the follicle. Follicles begin growth at a steady pace in a process known as folliculogenesis, and a small number enter the menstrual cycle. Menstruated oocytes continue meiosis I and arrest at meiosis II until fertilization. The process of meiosis in females occurs during oogenesis, and differs from the typical meiosis in that it features a long period of meiotic arrest known as the dictyate stage and lacks the assistance of centrosomes.

In males, meiosis occurs during spermatogenesis in the seminiferous tubules of the testicles. Meiosis during spermatogenesis is specific to a type of cell called spermatocytes that will later mature to become spermatozoa.

In female mammals, meiosis begins immediately after primordial germ cells migrate to the ovary in the embryo, but in the males, meiosis begins later, at the time of puberty. It is retinoic acid, derived from the primitive kidney (mesonephros) that stimulates meiosis in ovarian oogonia. Tissues of the male testis suppress meiosis by degrading retinoic acid, a stimulator of meiosis. This is overcome at puberty when cells within seminiferous tubules called Sertoli cells start making their own retinoic acid. Sensitivity to retinoic acid is also adjusted by proteins called nanos and DAZL.

MEIOSIS VS. MITOSIS

In order to understand meiosis, a comparison to mitosis is helpful. The table below shows the differences between meiosis and mitosis.

	Meiosis	**Mitosis**
End result	Normally four cells, each with half the number of chromosomes as the parent	Two cells, having the same number of chromosomes as the parent
Function	Production of gametes (sex cells) in sexually reproducing eukaryotes	Cellular reproduction, growth, repair, asexual reproduction
Where does it happen?	Reproductive cells of almost all eukaryotes (animals, plants, fungi, and protists)	All proliferating cells in all eukaryotes
Steps	Prophase I, Metaphase I, Anaphase I, Telophase I, Prophase II, Metaphase II, Anaphase II, Telophase II	Prophase, Prometaphase, Metaphase, Anaphase, Telophase
Genetically same as parent?	No	Yes
Crossing over happens?	Yes, normally occurs between each pair of homologous chromosomes	Very rarely
Pairing of homologous chromosomes?	Yes	No
Cytokinesis	Occurs in Telophase I and Telophase II	Occurs in Telophase
Centromeres split	Does not occur in Anaphase I, but occurs in Anaphase II	Occurs in Anaphase

G_1 PHASE

The **G_1 phase**, or **Gap 1 phase**, is the first of four phases of the cell cycle that takes place in eukaryotic cell division. In this part of interphase, the cell synthesizes mRNA and proteins in preparation for subsequent steps leading to mitosis. G1 phase ends when the cell moves into the S phase of interphase.

G1 phase together with the S phase and G2 phase comprise the long growth period of the cell cycle called interphase that takes place before cell division in mitosis (M phase).

During G_1 phase, the cell grows in size and synthesizes mRNA and proteins (Known as histones) that are required for DNA synthesis. Once the required proteins and growth are complete, the cell enters the next phase of the cell cycle, S phase. The duration of each phase, including the G_1phase, is different in many different types of cells. In human somatic cells, the cell cycle lasts about 18 hours, and the G1 phase takes up about 1/3 of that time. However, in Xenopus embryos, sea urchin embryos, and Drosophila embryos, the G1 phase is barely existent and is defined as the gap, if one exists, between the end of mitosis and the S phase.

G1 phase and the other subphases of the cell cycle may be affected by limiting growth factors such as nutrient supply, temperature, and room for growth. Sufficient nucleotides and amino acids must be present in order to synthesize mRNA and proteins. Physiological temperatures are optimal for cell growth. In humans, the normal physiological temperature is around 37 °C (98.6 °F).

G1 phase is particularly important in the cell cycle because it determines whether a cell commits to division or to leaving the cell cycle. If a cell is signaled to remain undivided, instead of moving onto the S phase, it will leave the G1 phase and move into a state of dormancy called the G0phase. Most nonproliferating vertebrate cells will enter the G0 phase.

Regulation of G1 Phase

Within the cell cycle, there is a stringent set of regulations known as the cell cycle control system that controls the timing and coordination of the phases to ensure a correct order of events. Biochemical triggers known as cyclin-dependent kinases (Cdks) switch on cell cycles events at the corrected time and in the correct order to prevent any mistakes.

There are three checkpoints in the cell cycle: the **G1/S Checkpoint** or the **Start checkpoint** in yeast; the G2/M checkpoint; and the spindle checkpoint.

Biochemical Regulators in G_1 Phase

During G1 phase, the G1/S cyclin activity rises significantly near the end of the G1 phase.

Complexes of cyclin that are active during other phases of the cell cycle are kept inactivated to prevent any cell-cycle events from occurring out of order. Three methods of preventing Cdk activity are found in G1 phase: inhibitory gene regulatory proteins suppress major cyclin genes from expression (*the way this is phrased it is unclear whether it is transcription of the gene or translation of the mRNA gene product that is being regulated); anaphase-promoting complex (APC) is activated, which targets and degrades S and M cyclins (but not G1/S cyclins); and a high concentration of Cdk inhibitors is found during G1 phase.

The Restriction Point

The Restriction Point **(R)** in the G1 phase is different from a checkpoint because it does not determine whether cell conditions are ideal to move on to the next phase, but it changes the course of the cell. After a vertebrate cell has been in the G1 phase for about three hours, the cell enters a restriction point in which it is decided whether the cell will move forward with the G1 phase or move into the dormant G0 phase.

This point also separates two halves of the G1 phase; the post-mitotic and pre-mitotic phases. Between the beginning of the G_1 phase (which is also after mitosis has occurred) and R, the cell is known as being in the G1-pm subphase, or the post-mitotic phase. After R and before S, the cell is known as being in G1-ps, or the pre S phase interval of the G1 phase.

In order for the cell to continue through the G1-pm, there must be a high amount of growth factors and a steady rate of protein synthesis, otherwise the cell will move into G0 phase.

Conflicting Research

Some authors will say that the Restriction Point and the G1/S Checkpoint are one and the same, but more recent studies have argued that there are two different points in the G_1 phase that check the progression of the cell. The first restriction point is growth-factor dependent and determines whether the cell moves into the G0 phase, while the second checkpoint is nutritionally-dependent and determines whether the cell moves into the S phase. Some of the confusion between researchers has been attributed to the study of mammalian vs. yeast cells.

The G1/S Checkpoint

The **G1/S checkpoint** is the point between G1 phase and the S phase in which the cell is cleared for progression into the S phase. Reasons the cell would not move into the S phase include insufficient cell growth, damaged DNA, or other preparations have not been completed.

At the G1/S checkpoint, formation of the G1/S cyclin with Cdk to form a complex commits the cell to a new division cycle. These complexes then activate S-Cdk complexes that move forward with DNA replication in the S phase. Concurrently, anaphase-promoting complex (APC) activity decreases significantly, allowing S and M cyclins to become activated.

If a cell does not clear to pass through to the S phase, it enters the dormant G0 phase in which there is no cellular growth or division.

The G1 Phase in Cancer

Many sources have linked irregularities in the G1 phase or the G1/S checkpoint to uncontrolled growth of tumors. In these cases where the G1phase is affected, it is generally because gene regulatory proteins of the E2F family have become unrestrained and increase G1/S cyclin gene expression, leading to uncontrolled cell-cycle entry.

However, the cure for some forms of cancer also lies in the G1 phase of the cell cycle. Many cancers including breast and skin cancers have been prevented from proliferating by causing the tumor cells to enter G1 cell cycle arrest, preventing the cells from dividing and spreading.

G1/S TRANSITION

The **G1/S transition** is a stage in the cell cycle at the boundary between the G1 phase, in which the cell grows, and the S phase, during which DNA is replicated.

During this transition, G1 cyclin D-cdk4/6 dimer phosphorylates retinoblastoma releasing transcription factor E2F, which then drives the transition from G1 to S phase. The G1/S transition is highly regulated by transcription factor p53 in order to halt the cell cycle when DNA is damaged.

It is a "point of no return" beyond which the cell is committed to dividing; in yeast this is called START and in multicellular eukaryotes it is termed the restriction point.

Cell Cycle Overview

The cell cycle is a process in which an ordered set of events leads to the growth and division into two daughter cells. The cell cycle is a cycle rather than a linear process because the two daughter cells produced repeat the cycle. This process contains two main phases, interphase, in which the cell grows and synthesizes a copy of its DNA, and the mitotic (M) phase, during which the cell separates its DNA and divides into two new daughter cells. Interphase is further broken down into the G1 (GAP 1) phase, S (Synthesis) phase, G2 (GAP 2) phase and the mitotic phase is broken down into mitosis and cytokinesis. During G1 phase, the cells that come out of mitosis grow larger and monitor the environment so that DNA can be replicated. Once the cells are large enough, the cell transitions into S phase and DNA is replicated in the nucleus. During S phase, the cell also duplicates the centrosome, or microtubule-organizing center, which is critical for DNA separation in the M phase. After complete synthesis of its DNA, the cell enters the G2 phase where it continues to grow in preparation for mitosis. Following interphase, the cell transitions into mitosis, containing four sub stages: prophase, anaphase, metaphase, and telophase. In mitosis, DNA condenses into chromosomes, which are lined up and separated by the mitotic spindle. After duplicate DNA is separated on opposite ends of the cell, the cytoplasm of the cell is split in two during cytokinesis resulting in two daughter cells.

Cell Cycle Regulation

As with most processes in the body, the cell cycle is highly regulated to prevent the synthesis of mutated cells and uncontrolled cell division that leads to tumor formation. The cell cycle control system is biochemically based so that the proteins of the mitosis promoting factor (MPF) control the transition from one phase to the next based on a series of checkpoints. MPF is a protein dimer made up of cyclin and cyclin-dependent kinase (Cdk), a serine and threonine kinase, which come together at different points in the cycle to control cell progression through the cycle. When cyclin binds to Cdk, Cdk becomes activated and phosphorylates serine and threonine on other proteins causing the activation and degradation of other proteins allowing the cell to transition through the cell cycle.

G1/Transition

In mid to late G1 phase, cyclin D bound to Cdk4/6 activates the expression of the S phase cyclin-Cdk components; however, the cell does not want S phase cyclins to become active in G1. Therefore, an inhibitor, protein Slc-1, is present that interacts with the dimer so that the S phase cyclin-Cdk dimer remains inactive until the cell is ready to move into S phase. After the cell has grown and is ready to synthesize DNA, G1 cyclin-Cdks phosphorylate the S phase cyclin inhibitor signaling ubiquination, resulting in the addition of groups to the inhibitor. Ubiquination of the inhibitor signals the SCF/proteasome to degrade the inhibitor releasing and allowing the S phase cyclin-Cdk to become activated and the cell moves into S phase. Once in S phase, cyclin-Cdks phosphorylate several factors on the replication complex promoting DNA replication by causing inhibitory proteins to fall off of replication complexes or through activation of components on the replication complex to induce DNA replication initiation.

Retinoblastoma and the G1/S Transition

Another dimer present during mid G1 is composed of retinoblastoma (Rb) protein and transcription factor E2F. When Rb is bound to E2F, E2F is inactive. As cyclin D is synthesized and activates Cdk4/6, the cylcin-Cdk targets Rb protein for phosphorylation. Upon phosphorylation, Rb changes conformation so that E2F is released and activated, binding to upstream regions of genes, initiating expression. Specifically, E2F drives the expression of other cyclins, including cyclin E and A, and genes necessary for DNA replication. Cyclin E either phosphorylates more Rb to further activate E2F and promote the expression of more Cyclin E, or it has the ability to increase expression of itself. Cyclin E also interacts with Cdk2 driving the cell cycle to progress from G1 to S phase.

The Role of Retinoblastoma in Tumor Formation

Retinoblastoma (Rb) is a cancer of the eye due to a mutant Rb protein. When Rb is mutated it becomes nonfunctional and is not able to inhibit the expression of transcription factor E2F. Therefore, E2F is always active and driving the cell cycle to progress from G1 to S phase. As a result, cell growth and division is unregulated causing tumor formation in the eye.

Cell cycle CHECKPOINTS

TO ENSURE proper cell division, the cell cycle utilizes numerous checkpoints to monitor cell progression and halt the cycle when processes go awry. These checkpoints include four DNA damage checkpoints, one unreplicated DNA checkpoint at the end of G2, one spindle assembly checkpoint in mitosis, and a chromosome segregation checkpoint during mitosis.

p53 as a regulator

Between G1 and S phase, three DNA damage checkpoints occur to ensure proper growth and synthesis of DNA prior to cell division. Damaged DNA during G1, before entry into S phase, and during S phase result in the expression of ATM/R protein. ATM/R protein then stabilizes and activates transcription factor p53 so that it can bind to upstream regions of genes, inducing the expression of proteins including p21CIP. p21CIP binds to and inhibits any cyclin-cdk present in the cell cycle, halting the cycle until DNA damage can be corrected.

Additional Processes at DNA Damage Checkpoints

Of the four DNA damage checkpoints, two have an additional process for monitoring DNA damage other than activating p53. Before entry into S phase and during S phase, ATM/R also activates Chk1/2 that inhibits Cdc25A, a protein responsible for activating cyclin-Cdk dimers. Without cyclin dimer activation, the cell cannot transition through the cycle. These two checkpoints have additional processes for regulation because replicating damaged DNA in S phase can be deleterious to the cell and more importantly, the organism.

G2 *PHASE*

G2 phase, or **Gap 2 phase**, is the second subphase of Interphase in the cell cycle directly preceding mitosis. It follows the successful completion of S phase, during which the cell's DNA is replicated. G2 phase ends with the onset of prophase, the first phase of mitosis in which the cell's chromatin condenses into chromosomes.

G_2 phase is a period of rapid cells growth and protein synthesis during which the cell readies itself for mitosis. Curiously, G2 phase is not a necessary part of the cell cycle, as some cell types (particularly young *Xenopus* embryos and some cancers) proceed directly from DNA replication to mitosis. Though much is known about the genetic network which regulates G2 phase and subsequent entry into mitosis, there is still much to be discovered concerning its significance and regulation, particularly in regards to cancer. One hypothesis is that the growth in G2 phase is regulated as a method of cell size control. Fission yeast (*Schizosaccharomyces pombe*) has been previously shown to employ such a mechanism, via Cdr2-mediated spatial regulation of Wee1 activity. Though Wee1 is a fairly conserved negative regulator of mitotic entry, no general mechanism of cell size control in G2 has yet been elucidated.

Biochemically, the end of G2 phase occurs when a threshold level of active cyclin B1/CDK1 complex, also known as Maturation promoting factor (MPF) has been reached. The activity of this complex is tightly regulated during G2. In particular, the G2 checkpoint arrests cells in G2 in response to DNA damage through inhibitory regulation of CDK1.

G_2/M Checkpoint

In vertebrate cells, the G_2/M DNA damage checkpoint consists of an arrest of the cell in G2 just before mitotic entry in response to genotoxic stress (such as UV radiation, oxidative stress, DNA intercalating agents, etc.) in both a p53-dependent and p53-independent manner. DNA damage signals cause activation of the transcription factor p53. CDK1 is directly inhibited by three transcriptional targets of p53: p21, Gadd45, and 14-3-3σ. Inactive Cyclin B1/CDK1 is sequestered in the nucleus by p21, while active Cyclin B1/CDK1 complexes are sequestered in the cytoplasm by 14-3-3σ. Gadd45 disrupts the binding of Cyclin B1 and CDK1 through direct interaction with CDK1. P53 also transcriptionally represses CDK1.

P53-independent G2 arrest is mainly affected through the actions of Chkl kinase. DNA damage is sensed by ATM and ATR (Rad3 and Mec1 in yeast), which then signal to Chk1 and Chk2. Chk1 then mediates the degradation of cdc25A, an activator of CDK1. ATR/ATM also activate p53, indicating that these pathways may act synergistically in regulating G2 arrest.

Both p53-dependent and p53-independent cell cycle arrest are not specific to G2; these same proteins function upstream in DNA damage checkpoints in G1 and S phase as well. In yeast, which has no p53 homolog, G2 arrest functions through the p53-independent pathway.

End of G2/Entry into Mitosis

Mitotic entry is determined by a threshold level of active cyclin B1/CDK1 complex. In vertebrates, there are five cyclin B isoforms (B1, B2, B3, B4, and B5), but specific role of each of these isoforms in regulating mitotic entry is still unclear. It is known that cyclin B1 can compensate for loss of both cyclin B2 (and vice versa in *Drosophila*). Cyclin B1/CDK1 activity is regulated both spatially and temporally during G_2 phase to ensure proper entry into mitosis.

Cyclin B1 transcription begins at the end of S phase after DNA replication. Its promoter contains consensus binding sequences for a number of transcription factors, including p53, p21, Ets, Ap-1, NF-Y, c-Myc, TFE3, and USF. Cyclin B1 accumulates in the cytoplasm throughout G2, where it binds to and activates CDK1's kinase activity. CDK1 activity is modulated primarily through regulation of its inhibitory phosphorylation sites at Thr14 and Tyr15. Weel and Mytl phosphorylate these two residues, with Weel acting on the Tyr15 site and Mytl acting predominantly on the Thr14 site. However, Mytl has a separate inhibitory effect on CDK1; it can also sequester CDK1 in the cytoplasm via interaction with Mytl's C-terminaldomain. CDK1 is dephosphorylated primarily through the actions of Cdc25, which can dephosphorylate both the Thr14 and Tyr15 residues of CDK1. There are three isoforms of Cdc25 (A, B, and C) in mammalian cells, all of which have been shown to have roles in regulation of G2 phase.

CDK1, in turn, phosphorylates and modulates the activity of Weel and the Cdc25 isoforms A and C. Specifically, CDK1 phosphorylation inhibits Weel kinase activity, activates Cdc25C phosphatase activity, and stabilizes Cdc25A. Thus, CDK1 forms a positive feedback loop with Cdc25 and a double negative feedback loop with Weel (essentially a net positive feedback loop). These loops encode a hysteretic bistable switch in CDK1 activity relative to Cyclin B1 levels. It is thought that this hysteretic behavior ensures that cells commit to mitosis even if cyclin B1 levels falter.

In mammals, cyclin B1/CDK1 translocation to the nucleus is activated by phosphorylation of five serine sites on cyclin B1's cytoplasmic retention site (CRS): S116, S26, S128, S133, and S147. In Xenopus laevis, Cyclin B1 contains four analogous CRS serine phosphorylation sites (S94, S96, S101, and S113) indicating that this mechanism is highly conserved. Nuclear export is also inactivated by phosphorylation of Cyclin B1's Nuclear export signal (NES). The regulators of these phosphorylation sites are still largely unknown but several factors have been identified, including extracellular signal-regulated kinases (ERKs), PLK1, and CDK1 itself. Upon reaching some threshold level of phosphorylation, translocation of cyclin B1/CDK1 to the nucleus is extremely rapid. Once in the nucleus, Cyclin B1/CDK1 phosphorylates many targets in preparation for mitosis, including Histone H1, nuclear lamins, centrosomal proteins, and Microtubule Associated Proteins (MAPs).

Recently, evidence has emerged suggesting a more important role for cyclin A2/CDK complexes in regulating entry into mitosis. Cyclin A2/CDK2 activity begins in early S phase and increases during G2. Cdc25B has been shown to dephosphorylate Tyr15 on CDK2 in early-to-mid G2 in a manner similar to the aforementioned CDK1 mechanism. Downregulation of cyclin A2 in U2OS cells increases Wee1 activity and lowers Plk1 and Cdc25C activity. However, cyclin A2/CDK complexes do not function strictly as activators of cyclin B1/CDK1 in G2, as CDK2 has been shown to be required for activation of the p53-independent G2 checkpoint activity, perhaps through a stabilizing phosphorylation on Cdc6.CDK2-/- cells also have aberrantly high levels of Cdc25A. Cyclin A2/CDK1 has also been shown to mediate proteosomal destruction of Cdc25B. These pathways are often deregulated in cancer.

BIOCHEMICAL SWITCHES IN THE CELL CYCLE

A series of **biochemical switches** control transitions between and within the various phases of the cell cycle. The cell cycle is a series of complex, ordered, sequential events that control how a single cell divides into two cells, and involves several different phases. The phases include the G1 and G2 phases, DNA replication or S phase, and the actual process of cell division, mitosis or M phase. During the M phase, the chromosomes separate and cytokinesis occurs.

The switches maintain the orderly progression of the cell cycle and act as checkpoints to ensure that each phase has been properly completed before progression to the next phase. For example, Cdk, or cyclin dependent kinase, is a major control switch for the cell cycle and it allows the cell to move from G1 to S or G2 to M by adding phosphate to protein substrates. Such multi-component (involving multiple inter-linked proteins) switches have been shown to generate decisive, robust (and potentially irreversible) transitions and trigger stable oscillations. As a result, they are a subject of active research that tries to understand how such complex properties are wired into biological control systems.

Feedback Loops

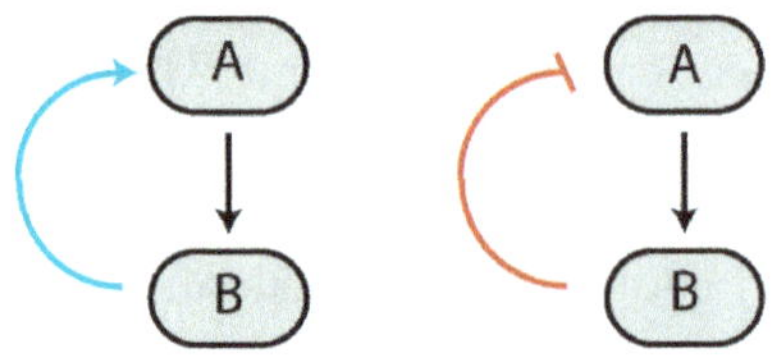

Many biological circuits produce complex outputs by exploiting one or more feedback loops. In a sequence of biochemical events, feedback would refer to a downstream element in the sequence (B in the image on the left) affecting some upstream component (A in the image on the left) to affect its own production or activation (output) in the future. If this element acts to enhance its own output, then it engages in positive feedback (blue arrow). A positive feedback loop is also known as a self-reinforcing loop, and it is possible that these loops can be part of a larger loop, as this is characteristic of regulatory circuits.

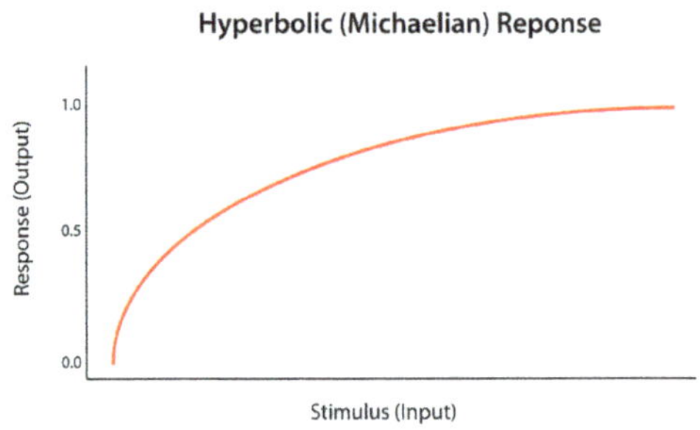

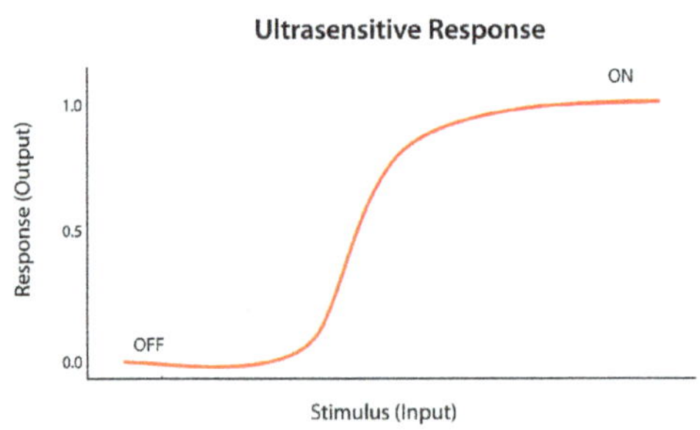

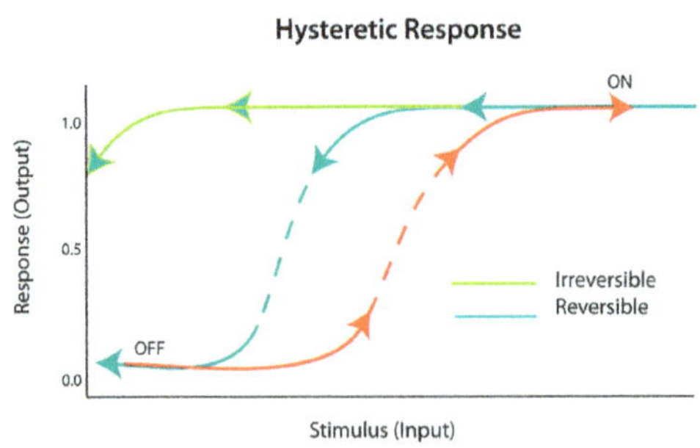

Conversely, if this element leads to its own inhibition through upstream elements, this is canonically negative feedback (red blunt arrow). A negative feedback loop is also known as a balancing loop, and it may be common to see oscillations in which a delayed negative feedback signal is used to maintain homeostatic balance in the system.

Feedback loops can be used for amplification (positive) or self-correction (negative). The right combination of positive and negative feedback loops can generate ultrasensitivity and bistability, which in turn can generate decisive transitions and oscillations.

Combination of Positive and Negative Feedback Loops

Positive and negative feedback loops do not always operate distinctly. In the mechanism of biochemical switches, they work together to create a flexible system. For example, according to Pfeuty & Kaneko (2009), to overcome a drawback in biochemical systems, positive feedback regulation loops may interact with negative regulation loops to facilitate escape from stable states. The coexistence of two stable states is known as bistability, which is often the result of positive feedback regulations.

An example that reveals the interaction of the multiple negative and positive feedback loops is the activation of cyclin-dependent protein kinases, or Cdks14. Positive feedback loops play a role by switching cells from low to high Cdk-activity. The interaction between the two types of loops is evident in mitosis. While positive feedback initiates mitosis, a negative feedback loop promotes the inactivation of the cyclin-dependent kinases by the anaphase-promoting complex. This example clearly shows the combined effects that positive and negative feedback loops have on cell-cycle regulation.

Ultrasensitivity

An "all-or-none" response to a stimulus is termed ultrasensitivity. In other words, a very small change in stimulus causes a very large change in response, producing a sigmoidal dose-response curve. An ultrasensitive response is described by the general equation $V = S^n/(S^n + K_m)$, known as the Hill equation, when n, the Hill coefficient, is more than 1. The steepness of the sigmoidal curve depends on the value of n. A value of $n = 1$ produces a hyperbolic or Michaelian response. Ultrasensitivity is achieved in a variety of systems; a notable example is the cooperative binding of the enzyme hemoglobin to its substrate. Since an ultrasensitive response is almost 'digital', it can be used to amplify a response to a stimulus or cause a decisive sharp transition (between 'off' and 'on' states).

Ultrasensitivity plays a large role in cell-cycle regulation. For example, Cdk1 and Wee1 are mitotic regulators, and they are able to inactivate each other through inhibitory phosphorylation. This represents a double negative feedback loop in which both regulators inactivate each other. According to Kim et al. (2007), there must be an ultrasensitive element to generate a bistable response. It turns out that Wee1 has an ultrasensitive response to Cdk1, and this likely arises because of substrate competition among the various phosphorylation sites on Wee1.

Bistability

Bistability implies hysteresis, and hysteresis implies multistability. Multistability indicates the presence of two or more stable states for a given input. Therefore, bistability is the ability of a system to exist in two steady states. In other words, there is a range of stimulus values for which the response can have two steady-state values. Bistability is accompanied by hysteresis, which means that the system approaches one of the two steady states preferentially depending on its history. Bistability requires feedback as well as an ultrasensitive circuit element.

Under the proper circumstances, positive and negative feedback loops can provide the conditions for bistability; for example, by having positive feedback coupled to an ultrasensitive response element with the circuit. A hysteretic bistable system can act as a robust reversible switch because it is harder for the system to transition between 'on' and 'off' states (compared to the equivalent monostable ultransensitive response). The system could also be poised such that one of the transitions is physically unattainable; for example, no amount of reduction in the stimulus will return the system to the 'off'-state once it is already in the 'on' state. This would form a robust irreversible switch.

There is no one-to-one correspondence between network topology, since many networks have a similar input and output relationship. A network topology does not imply input or output, and similarly input or output does not imply network topology. It is for this reason that parameterization is very important for circuit function. If the dynamics of the input are comparable or faster than the response of the system, the response may appear hysteretic.

Three cell cycle switches are described below that achieve abrupt and/or irreversible transitions by exploiting some of the mechanisms described above.

The G1/S Switch

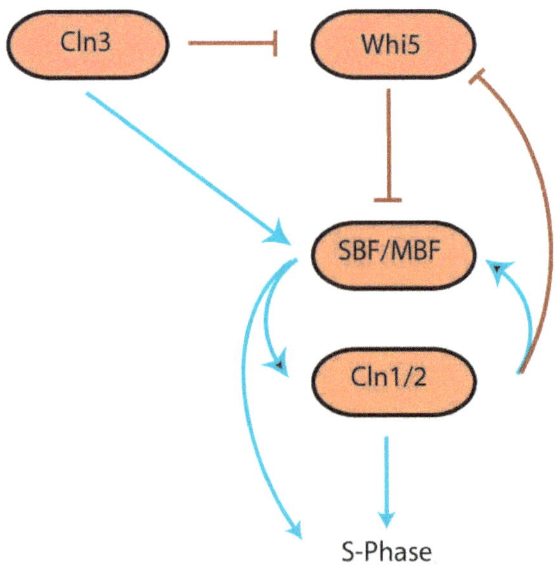

The G1/S transition, more commonly known as the Start checkpoint in budding yeast (the restriction point in other organisms) regulates cell cycle commitment. At this checkpoint, cells either arrest before DNA replication (due to limiting nutrients or a pheromone signal), prolong G1 (size control), or begin replication and progress through the rest of the cell cycle. The G1/S regulatory network or regulon in budding yeast includes the G1 cyclins Cln1, Cln2 and Cln3,

Cdc28 (Cdk1), the transcription factors SBF and MBF, and the transcriptional inhibitor Whi5. Cln3 interacts with Cdk1 to initiate the sequence of events by phosphorylating a large number of targets, including SBF, MBF and Whi5. Phosphorylation of Whi5 causes it to translocate out of the nucleus, preventing it from inhibiting SBF and MBF. Active SBF/MBF drive the G1/S transition by turning on the B-type cyclins and initiating DNA replication, bud formation and spindle body duplication. Moreover, SBF/MBF drives expression of Cln1 and Cln2, which can also interact with Cdk1 to promote phosphorylation of its targets.

This G1/S switch was initially thought to function as a linear sequence of events starting with Cln3 and ending in S phase. However, the observation that any one of the Clns was sufficient to activate the regulon indicated that Cln1 and Cln2 might be able to engage positive feedback to activate their own transcription. This would result in a continuously accelerating cycle that could act as an irreversible bistable trigger. Skotheim et al. used single-cell measurements in budding yeast to show that this positive feedback does indeed occur. A small amount of Cln3 induces Cln1/2 expression and then the feedback loop takes over, leading to rapid and abrupt exit of Whi5 from the nucleus and consequently coherent expression of G1/S regulon genes. In the absence of coherent gene expression, cells take longer to exit G1 and a significant fraction even arrest before S phase, highlighting the importance of positive feedback in sharpening the G1/S switch.

The G1/S cell cycle checkpoint controls the passage of eukaryotic cells from the first gap phase, G1, into the DNA synthesis phase, S. In this switch in mammalian cells, there are two cell cycle kinases that help to control the checkpoint: cell cycle kinases CDK4/6-cyclin D and CDK2-cyclin E.The transcription complex that includes Rb and E2F is important in controlling this checkpoint. In the first gap phase, the Rb-HDAC repressor complex binds to the E2F-DP1 transcription factors, therefore inhibiting the downstream transcription. The phosphorylation of Rb by CDK4/6 and CDK2 dissociates the Rb-repressor complex and serves as an on/off switch for the cell cycle. Once Rb is phosphorylated, the inhibition is released on the E2F transcriptional activity. This allows for the transcription of S phase genes encoding for proteins that amplify the G1 to S phase switch.

Many different stimuli apply checkpoint controls including TGFb, DNA damage, contact inhibition, replicative senescence, and growth factor withdrawal. The first four act by inducing members of the INK4 or Kip/Cip families of cell cycle kinase inhibitors. TGFb inhibits the transcription of Cdc25A, a phosphatase that activates the cell cycle kinases, and growth factor withdrawal activates GSK3b, which phosphorylates cyclin D. This leads to its rapid ubiquitination.

The G2/M switch

G2 is commenced by E2F-mediated transcription of cyclin A, which forms the cyclin A-Cdk2 complex. In order to proceed into mitosis, the cyclin B-Cdk1 complex (first discovered as MPF or M-phase promoting factor; Cdk1 is also known as Cdc2 in fission yeast and Cdc28 in budding yeast) is activated by Cdc25, a protein phosphatase. As mitosis starts, the nuclear envelope disintegrates, chromosomes condense and become visible, and the cell prepares for division. Cyclin B-Cdk1 activation results in nuclear envelope breakdown, which is a characteristic of the initiation of mitosis.

The cyclin B-Cdk1 complex participates in a regulatory circuit in which Cdk1 can phosphorylate and activate its activator, Cdc25 (positive feedback), and phosphorylate and inactivate its inactivator, the kinase Wee1 (double-negative feedback). This circuit could act as a bistable trigger with one stable steady state in G2 (Cdk1 and Cdc25 off, Wee1 on) and a second stable steady state in M phase (Cdk1 and Cdc25 active, Wee1 off). However, Wee1 is itself regulated by other factors, such as Cdr2.

Once cells are in mitosis, cyclin B-Cdk1 activates the anaphase-promoting complex (APC), which in turn inactivates cyclin B-Cdk1 by degrading cyclin B, eventually leading to exit from mitosis. Coupling the bistable Cdk1 response function to the negative feedback from the APC could generate what is known as a relaxation oscillator, with sharp spikes of Cdk1 activity triggering robust mitotic cycles. However, in a relaxation oscillator, the control parameter moves slowly relative to the system's response dynamics which may be an accurate representation of mitotic entry, but not necessarily mitotic exit.

It is necessary to inactivate the cyclin B-Cdk1 complex in order to exit the mitotic stage of the cell cycle. The cells can then return to the first gap phase G1 and wait until the cycle proceeds yet again.

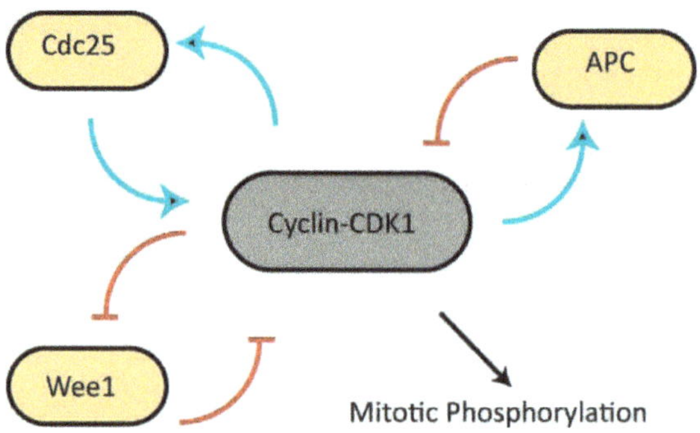

In 2003 Pomerening et al. provided strong evidence for this hypothesis by demonstrating hysteresis and bistability in the activation of Cdk1 in the cytoplasmic extracts of Xenopus oocytes. They first demonstrated a discontinuous sharp response of Cdk1 to changing concentrations of non-destructible Cyclin B (to decouple the Cdk1 response network from APC-mediated negative feedback). However, such a response would be consistent with both a monostable, ultransensitive transition and a bistable transition. To distinguish between these two possibilities, they measured the steady-state levels of active Cdk1 in response to changing cyclin levels, but in two separate experiments, one starting with an interphase extract and one starting with an extract already in mitosis. At intermediate concentrations of cyclin they found two steady-state concentrations of active Cdk1. Which of the two steady states was occupied depended on the history of the system, i.e.whether they started with interphase or mitotic extract, effectively demonstrating hysteresis and bistability.

In the same year, Sha et al. independently reached the same conclusion revealing the hysteretic loop also using Xenopus laevis egg extracts. In this article, three predictions of the Novak-Tyson model were tested in an effort to conclude that hysteresis is the driving force for "cell-cycle transitions into and out of mitosis". The predictions of the Novak-Tyson model are generic to all saddle-node bifurcations. Saddle-node bifurcations are extremely useful bifurcations in an imperfect world because they help describe biological systems which are not perfect. The first prediction was that the threshold concentration of cyclin to enter mitosis is higher than the threshold concentration of cyclin to exit mitosis, and this was confirmed by supplementing cycling egg extracts with non-degradable cyclin B and measuring the activation and inactivation threshold after the addition of cycloheximide (CHX), which is a protein synthesis inhibitor. Furthermore, the second prediction of the Novak-Tyson model was also validated: unreplicated deoxyribonucleic acid, or DNA, increases the threshold concentration of cyclin that is required to enter mitosis. In order to arrive at this conclusion, cytostatic factor released extracts were supplemented with CHX, APH (a DNA polymerase inhibitor), or both, and non-degradable cyclin B was added. The third and last prediction that was tested and proven true in this article was that the rate of Cdc2 activation slows down near the activation threshold concentration of cyclin. These predictions and experiments demonstrate the toggle-like switching behavior that can be described by hysteresis in a dynamical system.

Metaphase-Anaphase Switch

In the transition from metaphase to anaphase, it is crucial that sister chromatids are properly and simultaneously separated to opposite ends of the cell. Separation of sister-chromatids is initially strongly inhibited to prevent premature separation in late mitosis, but this inhibition is relieved through destruction of the inhibitory elements by the anaphase-promoting complex (APC) once sister-chromatid bi-orientation is achieved. One of these inhibitory elements is securin, which prevents the destruction of cohesin, the complex that holds the sister-chromatids together, by binding the protease separase which targets Scc1, a subunit of the cohesin complex, for destruction. In this system, the phosphatase Cdc14 can remove an inhibitory phosphate from securin, thereby facilitating the destruction of securin by the APC, releasing separase. As shown by Uhlmann et al., during the attachment of chromosomes to the mitotic spindle the chromatids remain paired because cohesion between the sisters prevents separation. Cohesion is established during DNA replication and depends on cohesin, which is a multisubunit complex composed of Scc1, Scc3, Smc2, and Smc3. In yeast at the metaphase-to-anaphase transition, Scc1 dissociates from the chromosomes and the sister chromatids separate. This action is controlled by the Esp1 protein, which is tightly bound by the anaphase inhibitor Pds1 that is destroyed by the anaphase-promoting complex. In order to verify that Esp1 does play a role in regulating Scc1 chromosome association, cell strains were arrested in G1 with an alpha factor. These cells stayed in arrest during the development. Esp1-1 mutant cells were used and the experiment was repeated, and Scc1 successfully bound to the chromosomes and remained associated even after the synthesis was terminated. This was crucial in showing that with Esp1, Scc1 is hindered in its ability to become stably associated with chromosomes during G1, and Esp1 can in fact directly remove Scc1 from chromosomes.

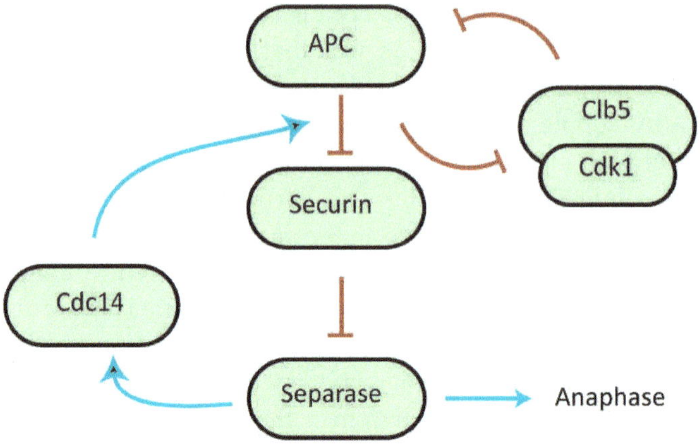

It has been shown by Holt et al. that separase activates Cdc14, which in turn acts on securin, thus creating a positive feedback loop that increases the sharpness of the metaphase to anaphase transition and coordination of sister-chromatid separation. Holt et al. probed the basis for the effect of positive feedback in securin phosphophorlyation by using mutant 'securin' strains of yeast, and testing how changes in the phosphoregulation of securin affects the synchrony of sister chromatid separation. Their results indicate that interfering with this positive securin-separase-cdc14 loop decreases sister chromatid separation synchrony. This positive feedback can hypothetically generate bistability in the transition to anaphase, causing the cell to make the irreversible decision to separate sister-chromatids.

Mitotic Exit

Mitotic Exit is an important transition point that signifies the end of mitosis and the onset of new G1 phase for a cell, and the cell needs to rely on specific control mechanisms to ensure that once it exits mitosis, it never returns to mitosis until it has gone through G1, S, and G2 phases and passed all the necessary checkpoints. Many factors including cyclins, cyclin-dependent kinases (CDKs), ubiquitin ligases, inhibitors of cyclin-dependent kinases, and reversible phosphorylations regulate mitotic exit to ensure that cell cycle events occur in correct order with least amount of errors. The end of mitosis is characterized by spindle breakdown, shortened kinetochore microtubules, and pronounced outgrowth of astral (non-kinetochore) microtubules. For a normal eukaryotic cell, mitotic exit is irreversible.

Proteolytic Degradation

Many speculations were made with regard to the control mechanisms employed by a cell to promote the irreversibility of mitotic exit in a eukaryotic model organism, the budding yeast *Saccharomyces cerevisiae*. Proteolytic degradation of cell cycle regulators and corresponding effects on the levels of cyclin-dependent kinases were proposed as a mechanism that promotes eukaryotic cell cycle and metaphase-to-anaphase transition in particular. In this theory, anaphase promoting complex (APC), a class of ubiquitin ligase, facilitates degradation of mitotic cyclins (Clb2) and anaphase-inhibiting factors (PDS1,

CUT2) to promote mitotic exit.APC ubiquitinates nine-amino acid motif known as the destruction box (D box) in the NH2-terminal domain of mitotic cyclins for degradation by proteasome. APC in association with Cdc20 (APC-Cdc20) ubiquitinates and targets mitotic cyclins (Clb2) for degradation at initial phase. Simultaneously, APC-Cdc20 mediates degradation of securins, which inhibit separases through binding, at anaphase onset. Released and active separase cleaves cohesin that held sister chromatids together, facilitating separation of sister chromatids and initiates mitotic exit by promoting release of Cdc14 from nucleolus.

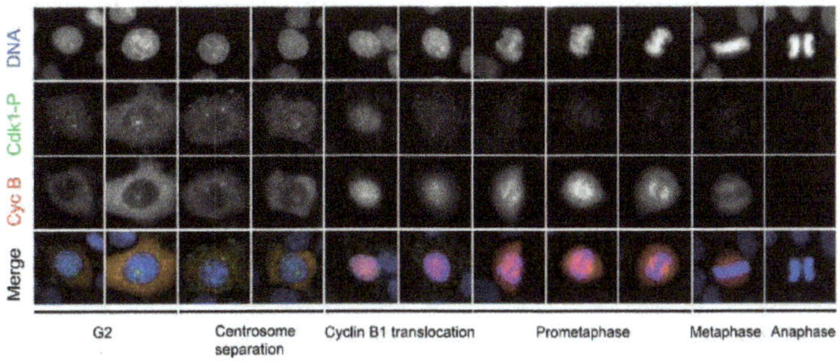

Fig. 1 Immunofluorescence patterns of cyclin B and phosphorylated cyclin-dependent kinase1 (Cdk1) in HeLa cells change as they go from G2 to anaphase.

At later phase, downregulation of Cdk1 and activation of Cdc14, a Cdh1-activating phosphatase, promotes formation of APC in association with Cdh1 (APC-Cdh1) to degrade Clb2s. Cdc20 and Cdh1, which are the activators of APC, recruit substrates such as securin and B-type cyclins(Clb) for ubiquitination. Without Cdk1-Clb2 complexes to phosphorylate proteins that are involved in spindle dynamics such as Sli15, Ase1, and Ask1, spindle elongation and chromosomal segregation are promoted, facilitating mitotic exit. The importance of proteolytic degradation in eukaryotic cell cycle changed the view of cell division as a simple kinase cascade to a more complex process in which interactions among phosphorylation, ubiquitination, and proteolysis are necessary. However, experiments using budding yeast cells with cdc28-as1, a INM-PP1 (ATP analog)-sensitive Cdk allele, proved that destruction of B-type cyclins (Clb) is not necessary for triggering irreversible mitotic exit. Clb2 degradation did shorten the Cdk1-inhibition period required for triggering irreversible mitotic exit indicating that cyclin proteolysis contributes to the dynamic nature of the eukaryotic cell cycle due to slower timescale of its action but is unlikely to be the major determining factor in triggering irreversible cell cycle transitions.

Sic1 levels

Discoveries were made which indicated the importance of the level of the inhibitors of cyclin-dependent kinases in regulating eukaryotic cell cycle. In particular, the level of Sic1, a stoichiometric inhibitor of Clb-CDK complexes in budding yeast, was shown to be particularly important in irreversible G1-S transition by irreversibly activating S phase kinases. Sic1 level was shown to play a major role in triggering irreversible mitotic exit (M-G1 transition) as well as in G1-S transition. During mitosis, decreasing levels of Cdk1 leads to the activation of Cdc14, a phosphatase that counteracts Cdk1 via activation of Cdh1 and Swi5, a transcriptional activator of Sic1 proteins. While degradation of Sic1 to a certain low level triggered the onset of S phase, accumulation of Sic1 to a certain high level was required to trigger irreversible mitotic exit. Cdk1-inhibitors could induce mitotic exit even when degradation of B-type cyclins was blocked by expression of non-degradable Clbs or proteasome inhibitors. However, sister chromatids failed to segregate, and cells reverted to mitosis once the inhibitors were washed away, indicating that a threshold level of the inhibitors needs to be achieved to trigger irreversible mitotic exit independently of cyclin degradations. Despite different thresholds of Sic1 level that are required to trigger mitotic exit compared to G1-S transition, the level of Sic1 was shown to play a key role in regulating eukaryotic cell cycle by inhibiting the activity of CDKs.

Dynamical Systems Approach

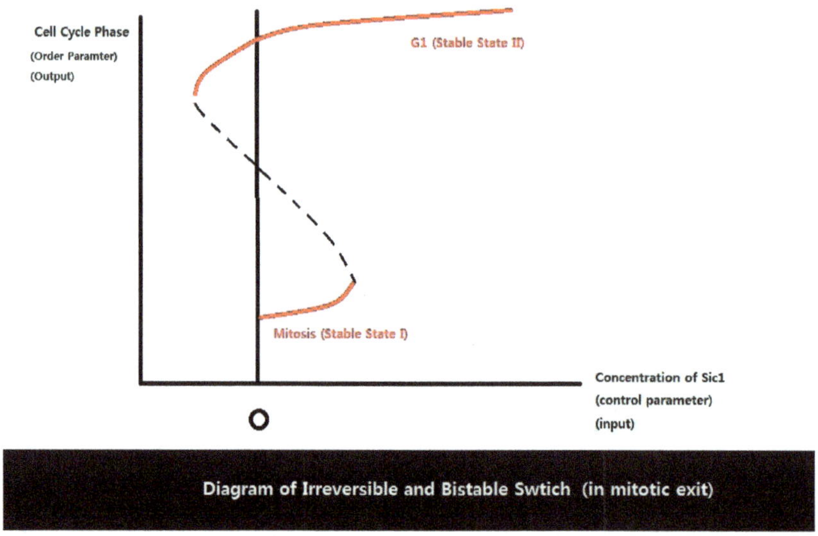

Fig. 2 Irreversible and bistable switch in mitotic exit with control parameter being Sic1 level and order parameter being cell cycle phases.

Because eukaryotic cell cycle involves a variety of proteins and regulatory interactions, dynamical systems approach can be taken to simplify a complex biological circuit into a general framework for better analysis. Among the four possible input/output relationships, the relationship between Sic1 level and mitotic exit seems to show the characteristics of an irreversible bistable switch, driven by feedback between APC-Cdh1, Sic1, and Clb2-Cdk1.Bistability is known to control biological functions such as cell cycle control and cellular differentiation and play a key role in many cellular regulatory networks. Bistable input/output relationship is characterized by two stable states with two bifurcation points. Multiple outputs are possible for one specific input in the region of bistability, marked by two bifurcation points. In addition, the bistable relationship displays hysteresis: the final state/output depends on the history of the input as well as the current value of input because the system has a memory. One bifurcation point has a negative control parameter value (the bifurcation point is on the other side of the axis), resulting in disconnection between the two stable states and irreversibility of the transition from one state to the other. With regard to mitotic exit, the two stable states are defined by mitosis and G1 phase. Once Sic1 level (input) accumulates beyond the threshold, irreversible transition occurs from mitosis (stable state I) to G1 phase (stable state II). In the imperfect environment, the only bifurcation that remains intact is saddle-node bifurcation. Saddle-node bifurcation does not break down (saddle-node is the expected generic behavior), while transcritical and pitchfork bifurcations break down in the presence of imperfections. Thus, the only one-dimensional bifurcation that can exist in imperfect biological world is the saddle-node bifurcation. The bistable relation between M-G1 transition and Sic1 level can be represented as a diagram of two saddle-node bifurcations in which the system's behavior changes qualitatively with a small change in control parameter, the amount of Sic1.

Systems-level Feedback

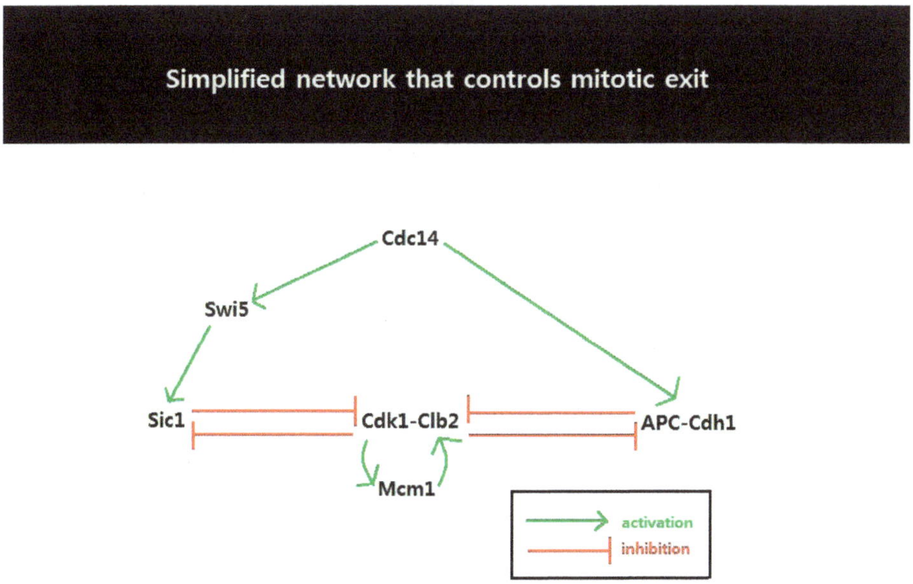

Fig. 3 Simplified network involving Cdk1-Clb2, APC-Cdh1, Sic1, and Cdc14. Double negative feedback loop, mediated by APC-Cdh1 and Sic1, is required to suppress Cdk1-Clb2 and trigger mitotic exit.

Because the behavior of cell cycle critically depends on the amount of Sic1 at the M-G1 transition state, the amount of Sic1 is tightly regulated by systems-level feedbacks. Because Cdk1-Clb2 inhibits Sic1 by phosphorylating Sic1 and making Sic1 available for degradation via ubiquitylation, APC-Cdh1-dependent degradation of Cdk1-Clb2 not only decreases the level of available Cdk1-Clb2 complexes but also increases the level of Sic1 which in turn further inhibits the function of Cdk1-Clb2. This activation of the double negative feedback loop is initiated from APC-Cdc20-dependent degradation of Cdk1-Clb2 and release of Cdc14 from nucleolar protein Net1/Cfi1. FEAR (Cdc14 early anaphase release) pathway facilitates Clb2-Cdk1-dependent phosphorylation of Net1 which transiently releases Cdc14 from Net1. The released Cdc14 and Clb2-Cdk1 complexes go onto form spindles that activates mitotic exit network (MEN). MEN allows sustained release of Cdc14 from the nucleolus, and Cdc14 counters the activity of Clb2-Cdk1 by activating Cdh1 and stabilizing Sic1 through activation of Sic1-transcriptional activator Swi5. Sic1 positively regulates itself by inhibiting Cdk1-Clb2 to release inhibition of Swi5, and Cdh1 also positively regulates itself by inhibiting Clb2-Cdk1 to release inhibition of MEN which can activate Cdc14 and subsequently Cdh1 itself. The double-negative feedback loop, formed by APC-Cdh1 and Sic1, is required to maintain low Clb2-Cdk1 activity because Clb2 auto-activates its synthesis by activating transcriptional factors, Fkh2–Mcm1 Ndd1 complex.

Implications

Eukaryotic cell cycle consists of various checkpoints and feedback loops to ensure faithful and successful cell division. During mitosis for example, when duplicated chromosomes are improperly attached to mitotic spindle, spindle assembly checkpoint (SAC) proteins including Mad and Bub inhibit APC-Cdc20 to delay entry into anaphase and B-type cyclin degradations. In addition, when mitotic spindles are misaligned, MEN and subsequently Cdc14 are inhibited in a Bub2 and Bfa1-dependent manner to prevent degradation of mitotic cyclins and anaphase entry. Sic1 is a nice example demonstrating how systems-level feedbacks interact to sense the environmental conditions and trigger cell cycle transitions. Even though actual M-G1 transition is vastly complex with numerous proteins and regulations involved, dynamical systems approach allows simplification of this complex system into bistable input/output relation with two saddle-node bifurcations in which the output (mitotic exit) depends on critical concentration of Sic1. Using one-dimensional analysis, it might be possible to explain many of the irreversible transition points in the eukaryotic cell cycle that are governed by systems-level control and feedback. Other examples of irreversible transition points include Start (irreversible commitment to a new cell division cycle) that can be explained by irreversible bistable switch whose control parameter is tightly regulated by the systemic feedbacks involving Cln2, Whi5, and SBF.

G0 *PHASE*

The **G0 phase**, or ***resting phase***' is a period in the cell cycle in which cells exist in an inactive, non-cycling state. The mammalian cell division cycle is controlled with extreme precision to complete the duplication of the genome and the segregation of replicated chromosome to daughter cells. In the past, cells were thought to enter G0 by default simply because of proliferation limits such as nutrient deprivation or contact inhibition. G0 is now viewed as a way for cells to preserve important functions over a long period.

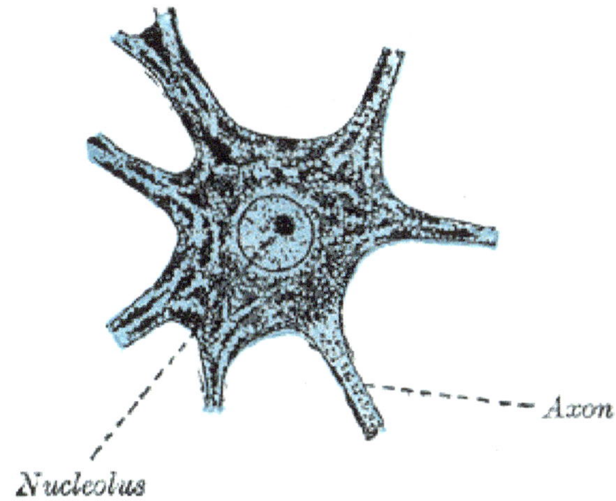

Many mammal cells, such as this 9x H neuron, remain permanently or semipermanently in G0.

G0 was first suggested as a cell state based on early cell cycle studies. When the first studies defined the four phases of the cell cycle using radioactive labeling techniques, it was discovered that not all cells in a population proliferate at similar rates. A population's "growth fraction" – or the fraction of the population that was growing – was actively proliferating, but other cells existed in a non-proliferative state. Some of these non-proliferating cells could respond to extrinsic stimuli and proliferate by re-entering the cell cycle. Early contrasting views either considered non-proliferating cells to simply be in an extended G1 phase or in a cell cycle phase distinct from G1 – termed G0. Subsequent research pointed to a restriction point (R-point) in G1where cells can enter G0 before the R-point but are committed to mitosis after the R-point. These early studies provided evidence for the existence of a G0 state to which access is restricted.

Diversity of G0 states

Three G_0 states exist and can be categorized as either reversible (quiescent) or irreversible (senescent and differentiated). Each of these three states can be entered from the G1 phase before the cell commits to the next round of the cell cycle. Quiescence refers to a reversible G_0 state where subpopulations of cells reside in a 'quiescent' state before entering the cell cycle after activation in response to extrinsic signals. Quiescent cells are often identified by low RNA content, lack of cell proliferation markers, and increased label retention indicating low cell turnover.Senescence is distinct from quiescence because senescence is an irreversible state that cells enter in response to DNA damage or degradation that would make a cell's progeny nonviable. Such DNA damage can occur from telomere shortening over many cell divisions as well as reactive oxygen species (ROS) exposure, oncogene activation, and cell-cell fusion. While senescent cells can no longer replicate, they remain able to perform many normal cellular functions. Senescence is often a biochemical alternative to the self-destruction of such a damaged cell by apoptosis. In contrast to cellular senescence, quiescence is not a reactive event but part of the core programming of several different cell types. Finally, differentiated cells are stem cells that have progressed through a differentiation program to reach a mature – terminally differentiated – state. Differentiated cells continue to stay in G0 and perform their main functions indefinitely.

Examples of Reversible G0 Phase

Tissue Stem Cells

Stem cells are cells with the unique ability to produce differentiated daughter cells and to preserve their stem cell identity through self-renewal. In mammals, most adult tissues contain tissue-specific stem cells that reside in the tissue and proliferate to maintain homeostasis for the lifespan of the organism. These cells can undergo immense proliferation in response to tissue damage before differentiating and engaging in regeneration. Some tissue stem cells exist in a reversible, quiescent state indefinitely until being activated by external stimuli. Many different types of tissue stem cells exist, including muscle stem cells (MuSCs), neural stem cells (NSCs), intestinal stem cells (ISCs), and many others.

Stem cell quiescence has been recently suggested to be composed of two distinct functional phases, G0 and an 'alert' phase termed G_{Alert}.Stem cells are believed to actively and reversibly transition between these phases to respond to injury stimuli and seem to gain enhanced tissue regenerative function in G_{Alert}. Thus, transition into G_{Alert} has been proposed as an adaptive response that enables stem cells to rapidly respond to injury or stress by priming them for cell cycle entry. In muscle stem cells, mTORC1 activity has been identified to control the transition from G0 into G_{Alert} along with signaling through the HGF receptor cMet.

Mature Hepatocytes

While a reversible quiescent state is perhaps most important for tissue stem cells to respond quickly to stimuli and maintain proper homeostasis and regeneration, reversible G_0 phases can be found in non-stem cells such as mature hepatocytes. Hepatocytes are typically quiescent in normal livers but undergo limited replication (less than 2 cell divisions) during liver regeneration after partial hepatectomy. However, in certain cases, hepatocytes can experience immense proliferation (more than 70 cell divisions) indicating that their proliferation capacity is not hampered by existing in a reversible quiescent state.

Examples of Irreversible G0 phase

Senescent Cells

Often associated with aging and age-related diseases in vivo, senescent cells can be found in many renewable tissues, including the stroma, vasculature, hematopoietic system, and many epithelial organs. Resulting from accumulation over many cell divisions, senescence is often seen in age-associated degenerative phenotypes. Senescent fibroblasts in models of breast epithelial cell function have been found to disrupt milk protein production due to secretion of matrix metalloproteinases. Similarly, senescent pulmonary artery smooth muscle cells caused nearby smooth muscle cells to proliferate and migrate, perhaps contributing to hypertrophy of pulmonary arteries and eventually pulmonary hypertension.

Differentiated Muscle

During skeletal myogenesis, cycling progenitor cells known as myoblasts differentiate and fuse together into non-cycling muscle cells called myocytes that remain in a terminal G0 phase. As a result, the fibers that make up skeletal muscle (myofibers) are cells with multiple nuclei, referred to as myonuclei, since each myonuclei originated from a single myoblast. Skeletal muscle cells continue indefinitely to provide contractile force through simultaneous contractions of cellular structures called sarcomeres. Importantly, these cells are kept in a terminal G0 phase since disruption of muscle fiber structure after myofiber formation would prevent proper transmission of force through the length of the muscle. Muscle growth can be stimulated by growth or injury and involves the recruitment of muscle stem cells – also known as satellite cells – out of a reversible quiescent state. These stem cells differentiate and fuse to generate new muscle fibers both in parallel and in series to increase force generation capacity.

Cardiac muscle is also formed through myogenesis but instead of recruiting stem cells to fuse and form new cells, heart muscle cells – known as cardiomyocytes – simply increase in size as the heart grows larger. Similarly to skeletal muscle, if cardiomyocytes had to continue dividing to add muscle tissue the contractile structures necessary for heart function would be disrupted.

Differentiated Bone

Of the four major types of bone cells, osteocytes are the most common and also exist in a terminal G_0 phase. Osteocytes arise from osteoblasts that are trapped within a self-secreted matrix. While osteocytes also have reduced synthetic activity, they still serve bone functions besides generating structure. Osteocytes work through various mechanosensory mechanisms to assist in the routine turnover over bony matrix.

Differentiated Nerve

Outside of a few neurogenic niches in the brain, most neurons are fully differentiated and reside in a terminal G0 phase. These fully differentiated neurons form synapses where electrical signals are transmitted by axons to the dendrites of nearby neurons. In this G0 state, neurons continue functioning until senescence or apoptosis.

APOPTOSIS

Apoptosis is a process of programmed cell death that occurs in multicellular organisms. Biochemical events lead to characteristic cell changes (morphology) and death. These changes include blebbing, cell shrinkage, nuclear fragmentation, chromatin condensation, chromosomal DNA fragmentation, and global mRNA decay. Between 50 and 70 billion cells die each day due to apoptosis in the average human adult. For an average child between the ages of 8 and 14, approximately 20 billion to 30 billion cells die a day.

In contrast to necrosis, which is a form of traumatic cell death that results from acute cellular injury, apoptosis is a highly regulated and controlled process that confers advantages during an organism's lifecycle. For example, the separation of fingers and toes in a developing human embryo occurs because cells between the digits undergo apoptosis. Unlike necrosis, apoptosis produces cell fragments called apoptotic bodies that phagocytic cells are able to engulf and quickly remove before the contents of the cell can spill out onto surrounding cells and cause damage to the neighboring cells.

Because apoptosis cannot stop once it has begun, it is a highly regulated process. Apoptosis can be initiated through one of two pathways. In the *intrinsic pathway* the cell kills itself because it senses cell stress, while in the *extrinsic pathway* the cell kills itself because of signals from other cells. Both pathways induce cell death by activating caspases, which are proteases, or enzymes that degrade proteins. The two pathways both activate initiator caspases, which then activate executioner caspases, which then kill the cell by degrading proteins indiscriminately.

Research on apoptosis has increased substantially since the early 1990s. In addition to its importance as a biological phenomenon, defective apoptotic processes have been implicated in a wide variety of diseases. Excessive apoptosis causes atrophy, whereas an insufficient amount results in uncontrolled cell proliferation, such as cancer. Some factors like Fas receptors and caspases promote apoptosis, while some members of the Bcl-2 family of proteins inhibit apoptosis.

Discovery and Etymology

German scientist Karl Vogt was first to describe the principle of apoptosis in 1842. In 1885, anatomist Walther Flemming delivered a more precise description of the process of programmed cell death. However, it was not until 1965 that the topic was resurrected. While studying tissues using electron microscopy, John Foxton Ross Kerr at University of Queensland was able to distinguish apoptosis from traumatic cell death. Following the publication of a paper describing the phenomenon, Kerr was invited to join Alastair R Currie, as well as Andrew Wyllie, who was Curries graduate student, at University of Aberdeen. In 1972, the trio published a seminal article in the British Journal of Cancer. Kerr had initially used the term programmed cell necrosis, but in the article, the process of natural cell death was called *apoptosis*. Kerr, Wyllie and Currie credited James Cormack, a professor of Greek language at University of Aberdeen, with suggesting the term apoptosis. Kerr received the Paul Ehrlich and Ludwig Darmstaedter Prize on March 14, 2000, for his description of apoptosis. He shared the prize with Boston biologist H. Robert Horvitz.

For many years, the terms "apoptosis" and "programmed cell death" were not highly cited. What transformed cell death from obscurity to a major field of research were two things: the identification of components of the cell death control and effector mechanisms, and the linkage of abnormalities in cell death to human disease, in particular cancer.

The 2002 Nobel Prize in Medicine was awarded to Sydney Brenner, Horvitz and John E. Sulston for their work identifying genes that control apoptosis. The genes were identified by studies in the nematode *C. elegans* and these same genes function in humans for apoptosis.

In Greek, apoptosis translates to the falling off of leaves from a tree. Cormack, professor of Greek language, reintroduced the term for medical use as it had a medical meaning for the Greeks over two thousand years before. Hippocrates used the term to mean "the falling off of the bones". Galen extended its meaning to "the dropping of the scabs". Cormack was no doubt aware of this usage when he suggested the name. Debate continues over the correct pronunciation, with opinion divided between a pronunciation with the second *p* silent and the second *p* pronounced , as in the original Greek. In English, the *p* of the Greek *-pt-* consonant cluster is typically silent at the beginning of a word (e.g. pterodactyl, Ptolemy), but articulated when used in combining forms preceded by a vowel, as in helicopter or the orders of insects: diptera, lepidoptera, etc.

In the original Kerr, Wyllie & Currie paper, there is a footnote regarding the pronunciation:

We are most grateful to Professor James Cormack of the Department of Greek, University of Aberdeen, for suggesting this term. The word apoptosis is used in Greek to describe the dropping off or falling off of petals from flowers, or leaves from trees. To show the derivation clearly, we propose that the stress should be on the penultimate syllable, the second half of the word being pronounced like ptosis (with the p silent), which comes from the same root to fall, and is already used to describe the drooping of the upper eyelid.

Activation Mechanisms

A cell initiates intracellular apoptotic signaling in response to a stress, which may bring about cell suicide. The binding of nuclear receptors by glucocorticoids, heat, radiation, nutrient deprivation, viral infection, hypoxia and increased intracellular calcium concentration, for example, by damage to the membrane, can all trigger the release of intracellular apoptotic signals by a damaged cell. A number of cellular components, such as poly ADP ribose polymerase, may also help regulate apoptosis.

Before the actual process of cell death is precipitated by enzymes, apoptotic signals must cause regulatory proteins to initiate the apoptosis pathway. This step allows those signals to cause cell death, or the process to be stopped, should the cell no longer need to die. Several proteins are involved, but two main methods of regulation have been identified: the targeting of mitochondria functionality, or directly transducing the signal via adaptor proteins to the apoptotic mechanisms. An extrinsic pathway for initiation identified in several toxin studies is an increase in calcium concentration within a cell caused by drug activity, which also can cause apoptosis via a calcium binding protease calpain.

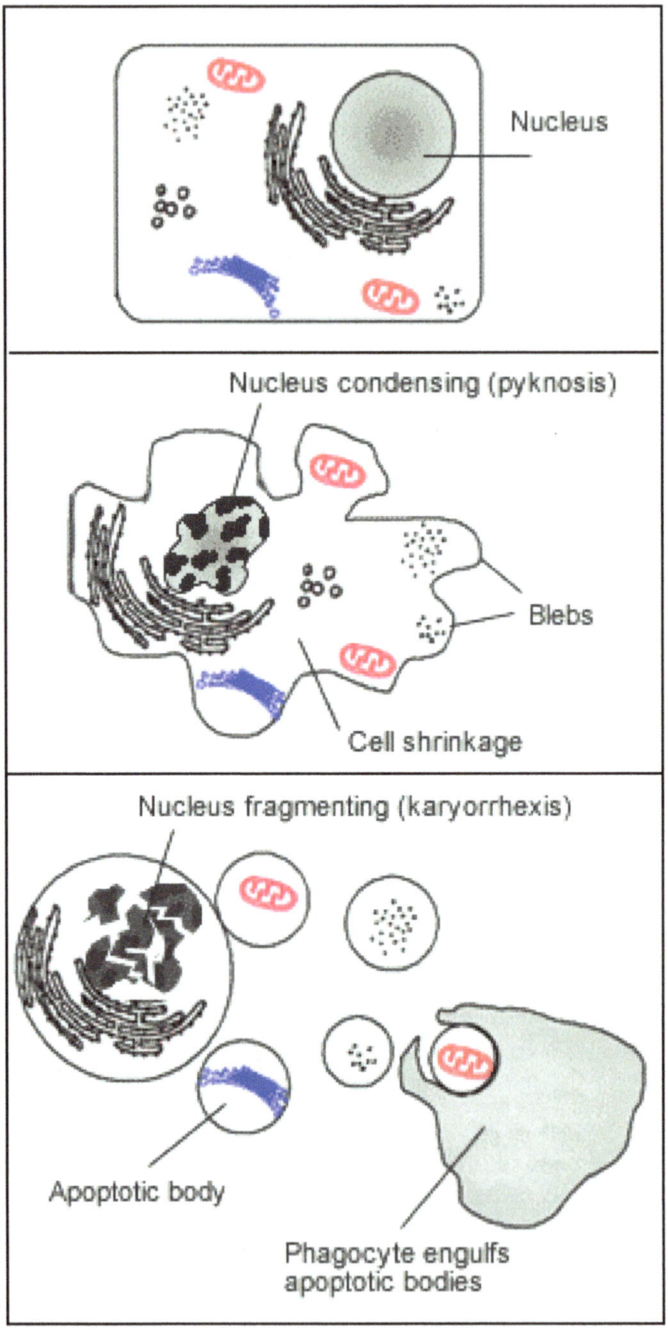

The initiation of apoptosis is tightly regulated by activation mechanisms, because once apoptosis has begun, it inevitably leads to the death of the cell. The two best-understood activation mechanisms are the intrinsic pathway (also called the mitochondrial pathway) and the extrinsic pathway. The *intrinsic pathway* is activated by intracellular signals generated when cells are stressed and depends on the release of proteins from the intermembrane space of mitochondria. The *extrinsic pathway* is activated by extracellular ligands binding to cell-surface death receptors, which leads to the formation of the death-inducing signaling complex (DISC).

Intrinsic Pathway

The mitochondria are essential to multicellular life. Without them, a cell ceases to respire aerobically and quickly dies. This fact forms the basis for some apoptotic pathways. Apoptotic proteins that target mitochondria affect them in different ways. They may cause mitochondrial swelling through the formation of membrane pores, or they may increase the permeability of the mitochondrial membrane and cause apoptotic effectors to leak out. They are very closely related to intrinsic pathway, and tumors arise more frequently through intrinsic pathway than the extrinsic pathway because of sensitivity. There is also a growing body of evidence indicating that nitric oxide is able to induce apoptosis by helping to dissipate the membrane potential of mitochondria and therefore make it more permeable. Nitric oxide has been implicated in initiating and inhibiting apoptosis through its possible action as a signal molecule of subsequent pathways that activate apoptosis.

Mitochondrial proteins known as SMACs (second mitochondria-derived activator of caspases) are released into the cell's cytosol following the increase in permeability of the mitochondia membranes. SMAC binds to *proteins that inhibit apoptosis* (IAPs) thereby deactivating them, and preventing the IAPs from arresting the process and therefore allowing apoptosis to proceed. IAP also normally suppresses the activity of a group of cysteine proteases called caspases, which carry out the degradation of the cell. Therefore, the actual degradation enzymes can be seen to be indirectly regulated by mitochondrial permeability.

Cytochrome c is also released from mitochondria due to formation of a channel, the mitochondrial apoptosis-induced channel (MAC), in the outer mitochondrial membrane, and serves a regulatory function as it precedes morphological change associated with apoptosis. Once cytochrome c is released it binds with Apoptotic protease activating factor – 1 (*Apaf-1*) and ATP, which then bind to *pro-caspase-9* to create a protein complex known as an apoptosome. The apoptosome cleaves the pro-caspase to its active form of caspase-9, which in turn activates the effector *caspase-3*.

MAC (not to be confused with the Membrane Attack Complex formed by complement activation, also commonly denoted as MAC), also called "Mitochondrial Outer Membrane Permeabilization Pore" is regulated by various proteins, such as those encoded by the mammalian Bcl-2 family of anti-apoptopic genes, the homologs of the *ced-9* gene found in *C. elegans*. *Bcl-2* proteins are able to promote or inhibit apoptosis by direct action on MAC/MOMPP. Bax and/or Bak form the pore, while Bcl-2, Bcl-xL or Mcl-1 inhibit its formation.

Extrinsic Pathway

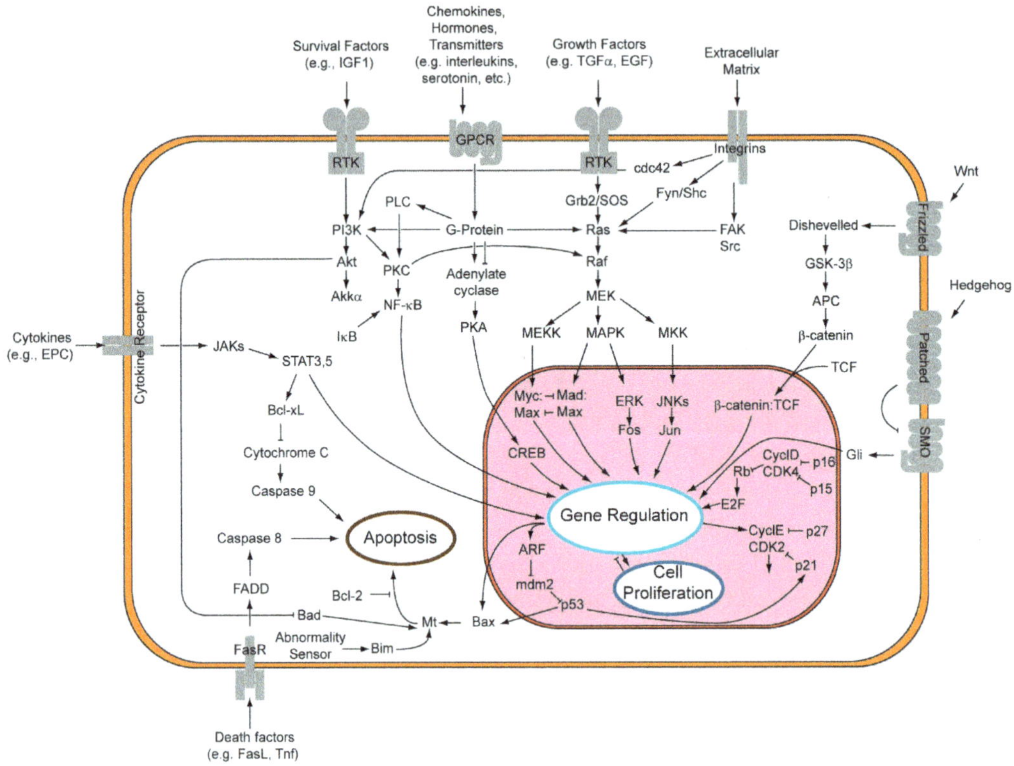

Overview of signal transduction pathways.

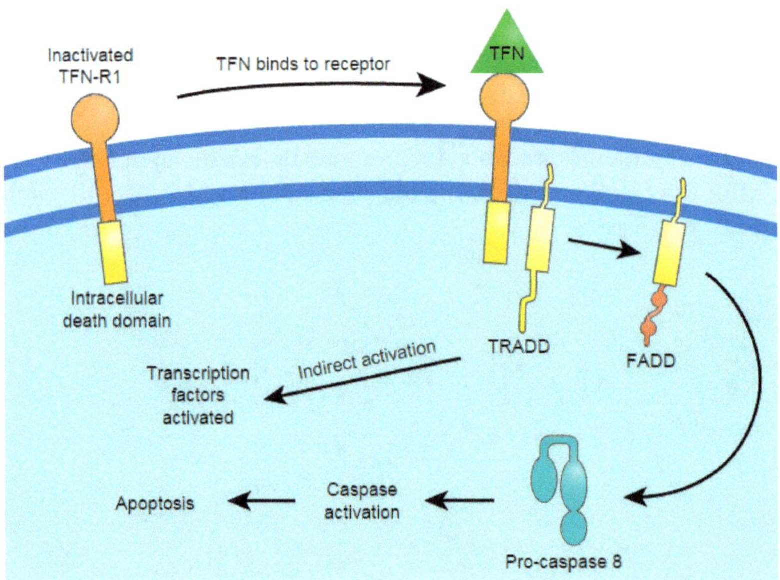

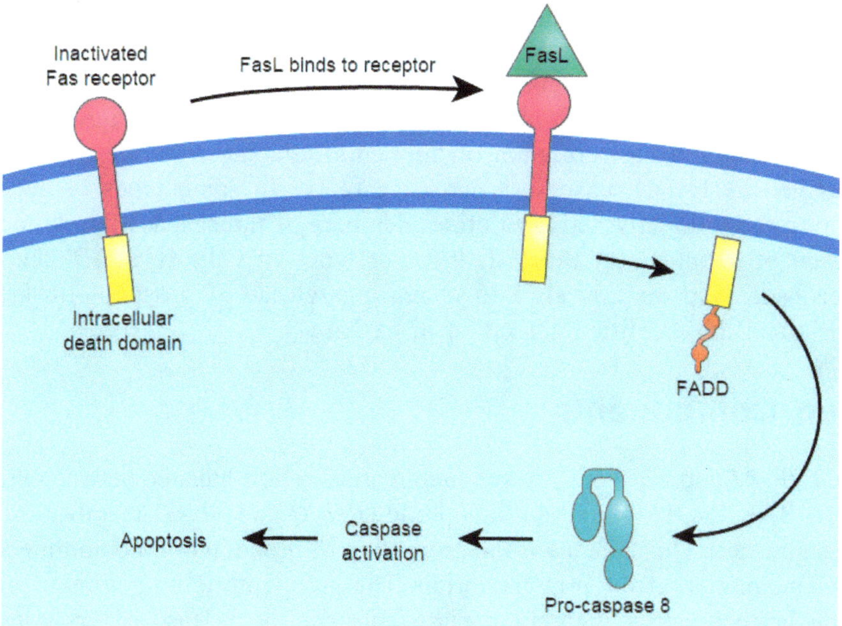

Overview of TNF (left) and Fas (right) signalling in apoptosis, an example of direct signal transduction.

Two theories of the direct initiation of apoptotic mechanisms in mammals have been suggested: the *TNF-induced* (tumour necrosis factor) model and the *Fas-Fas ligand-mediated* model, both involving receptors of the *TNF receptor* (TNFR) family coupled to extrinsic signals.

TNF path

TNF-alpha is a cytokine produced mainly by activated macrophages, and is the major extrinsic mediator of apoptosis. Most cells in the human body have two receptors for TNF-alpha: TNFR1 and TNFR2. The binding of TNF-alpha to TNFR1 has been shown to initiate the pathway that leads to caspase activation via the intermediate membrane proteins TNF receptor-associated death domain (TRADD) and Fas-associated death domain protein (FADD). cIAP1/2 can inhibit TNF-α signaling by binding to TRAF2. FLIP inhibits the activation of caspase-8. Binding of this receptor can also indirectly lead to the activation of transcription factors involved in cell survival and inflammatory responses. However, signalling through TNFR1 might also induce apoptosis in a caspase-independent manner. The link between TNF-alpha and apoptosis shows why an abnormal production of TNF-alpha plays a fundamental role in several human diseases, especially in autoimmune diseases.

Fas path

The fas receptor (First apoptosis signal) – (also known as *Apo-1* or *CD95*) is a transmembrane protein of the TNF family which binds the Fas ligand (FasL). The interaction between Fas and FasL results in the formation of the *death-inducing signaling complex* (DISC), which contains the FADD, caspase-8 and caspase-10. In some types of cells (type I), processed caspase-8 directly activates other members of the caspase family, and triggers the execution of apoptosis of the cell. In other types of cells (type II), the *Fas*-DISC starts a feedback loop that spirals into increasing release of proapoptotic factors from mitochondria and the amplified activation of caspase-8.

Common Components

Following *TNF-R1* and *Fas* activation in mammalian cells a balance between proapoptotic (BAX, BID, BAK, or BAD) and anti-apoptotic (*Bcl-Xl* and *Bcl-2*) members of the *Bcl-2* family is established. This balance is the proportion of proapoptotic homodimers that form in the outer-membrane of the mitochondrion. The proapoptotic homodimers are required to make the mitochondrial membrane permeable for the release of caspase activators such as cytochrome c and SMAC. Control of proapoptotic proteins under normal cell conditions of nonapoptotic cells is incompletely understood, but in general, Bax or Bak are activated by the activation of BH3-only proteins, part of the Bcl-2 family.

Caspases Caspases play the central role in the transduction of ER apoptotic signals. Caspases are proteins that are highly conserved, cysteine-dependent aspartate-specific proteases. There are two types of caspases: initiator caspases, caspase 2,8,9,10,11,12, and effector caspases, caspase 3,6,7. The activation of initiator caspases requires binding to specific oligomeric activator protein. Effector caspases are then activated by these active initiator caspases through proteolytic cleavage. The active effector caspases then proteolytically degrade a host of intracellular proteins to carry out the cell death program.

Caspase-independent apoptotic pathway There also exists a caspase-independent apoptotic pathway that is mediated by AIF (apoptosis-inducing factor).

Apoptosis Model in Amphibians

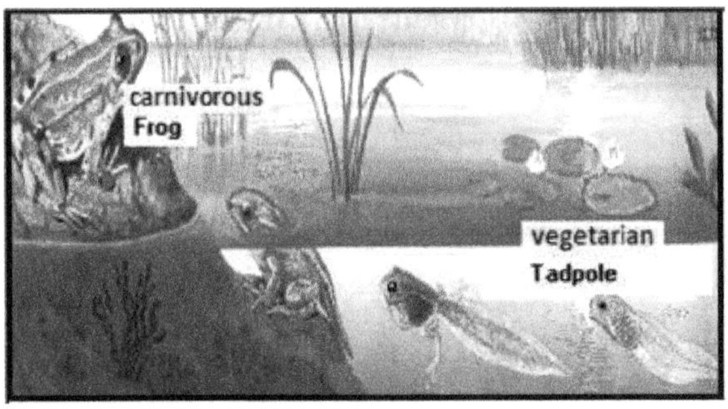

Amphibian Metamorphosis.

Amphibian frog *Xenopus laevis* serves as an ideal model system for the study of the mechanisms of apoptosis. In fact, iodine and thyroxine also stimulate the spectacular apoptosis of the cells of the larval gills, tail and fins in amphibians metamorphosis, and stimulate the evolution of their nervous system transforming the aquatic, vegetarian tadpole into the terrestrial, carnivorous frog.

Proteolytic Caspase Cascade: Killing the Cell

Many pathways and signals lead to apoptosis, but these converge on a single mechanism that actually causes the death of the cell. After a cell receives stimulus, it undergoes organized degradation of cellular organelles by activated proteolytic caspases. In addition to the destruction of cellular organelles, mRNA is rapidly and globally degraded by a mechanism that is not yet fully characterized. mRNA decay is triggered very early in apoptosis. A cell undergoing apoptosis shows a characteristic morphology:

- Cell shrinkage and rounding are shown because of the breakdown of the proteinaceous cytoskeleton by caspases.
- The cytoplasm appears dense, and the organelles appear tightly packed.
- Chromatin undergoes condensation into compact patches against the nuclear envelope (also known as the perinuclear envelope) in a process known as pyknosis, a hallmark of apoptosis.
- The nuclear envelope becomes discontinuous and the DNA inside it is fragmented in a process referred to as karyorrhexis. The nucleus breaks into several discrete *chromatin bodies* or *nucleosomal units* due to the degradation of DNA.
- The cell membrane shows irregular buds known as blebs.
- The cell breaks apart into multiple vesicles called *apoptotic bodies*, which are then phagocytosed.

Apoptosis progresses quickly and its products are quickly removed, making it difficult to detect or visualize. During karyorrhexis, endonuclease activation leaves short DNA fragments, regularly spaced in size. These give a characteristic "laddered" appearance on agar gel after electrophoresis. Tests for DNA laddering differentiate apoptosis from ischemic or toxic cell death.

Removal of Dead Cells

The removal of dead cells by neighboring phagocytic cells has been termed efferocytosis. Dying cells that undergo the final stages of apoptosis display phagocytotic molecules, such as phosphatidylserine, on their cell surface.Phosphatidylserine is normally found on the inner leaflet surface of the plasma membrane, but is redistributed during apoptosis to the extracellular surface by a protein known as scramblase. These molecules mark the cell for phagocytosis by cells possessing the appropriate receptors, such as macrophages. The removal of dying cells by phagocytes occurs in an orderly manner without eliciting an inflammatory response.

Pathway Knock-Outs

Many knock-outs have been made in the apoptosis pathways to test the function of each of the proteins. Several caspases, in addition to APAF-1 and FADD, have been mutated to determine the new phenotype. In order to create a tumor necrosis factor (TNF) knockout, an exon containing the nucleotides 3704-5364 was removed from the gene. This exon encodes a portion of the mature TNF domain, as well as the leader sequence, which is a highly conserved region necessary for proper intracellular processing. TNF-/- mice develop normally and have no gross structural or morphological abnormalities. However, upon immunization with SRBC (sheep red blood cells), these mice demonstrated a deficiency in the maturation of an antibody response; they were able to generate normal levels of IgM, but could not develop specific IgG levels. Apaf-1 is the protein that turns on caspase 9 by cleavage to begin the caspase cascade that leads to apoptosis. Since a -/- mutation in the APAF-1 gene is embryonic lethal, a gene trap strategy was used in order to generate an APAF-1 -/- mouse. This assay is used to disrupt gene function by creating an intragenic gene fusion. When an APAF-1 gene trap is introduced into cells, many morphological changes occur, such as spina bifida, the persistence of interdigital webs, and open brain. In addition, after embryonic day 12.5, the brain of the embryos showed several structural changes. APAF-1 cells are protected from apoptosis stimuli such as irradiation. A BAX-1 knock-out mouse exhibits normal forebrain formation and a decreased programmed cell death in some neuronal populations and in the spinal cord, leading to an increase in motor neurons.

The caspase proteins are integral parts of the apoptosis pathway, so it follows that knock-outs made have varying damaging results. A caspase 9 knock-out leads to a severe brain malformation. A caspase 8 knock-out leads to cardiac failure and thus embryonic lethality. However, with the use of cre-lox technology, a caspase 8 knock-out has been created that exhibits an increase in peripheral T cells, an impaired T cell response, and a defect in neural tube closure. These mice were found to be resistant to apoptosis mediated by CD95, TNFR, etc. but not resistant to apoptosis caused by UV irradiation, chemotherapeutic drugs, and other stimuli. Finally, a caspase 3 knock-out was characterized by ectopic cell masses in the brain and abnormal apoptotic features such as membrane blebbing or nuclear fragmentation. A remarkable feature of these KO mice is that they have a very restricted phenotype: Casp3, 9, APAF-1 KO mice have deformations of neural tissue and FADD and Casp 8 KO showed defective heart development, however in both types of KO other organs developed normally and some cell types were still sensitive to apoptotic stimuli suggesting that unknown proapoptotic pathways exist.

Methods for Distinguishing Apoptotic from Necrotic (Necroptotic) Cells

In order to perform analysis of apoptotic versus necrotic (necroptotic) cells, one can do analysis of morphology by time-lapse microscopy, flow fluorocytometry, and transmission electron microscopy. There are also various biochemical techniques for analysis of cell surface markers (phosphatidylserine exposure versus cell permeability by flow fluorocytometry), cellular markers such as DNA fragmentation (flow fluorocytometry), caspase activation, Bid cleavage, and cytochrome c release (Western blotting). It is important to know how primary and secondary necrotic cells can be distinguished by analysis of supernatant for caspases, HMGB1, and release of cytokeratin 18. However, no distinct surface or biochemical markers of necrotic cell death have been identified yet, and only negative markers are available. These include absence of apoptotic markers (caspase activation, cytochrome c release, and oligonucleosomal DNA fragmentation) and differential kinetics of cell death markers (phosphatidylserine exposure and cell membrane permeabilization). A selection of techniques that can be used to distinguish apoptosis from necroptotic cells could be found in these references.

Implication in Disease

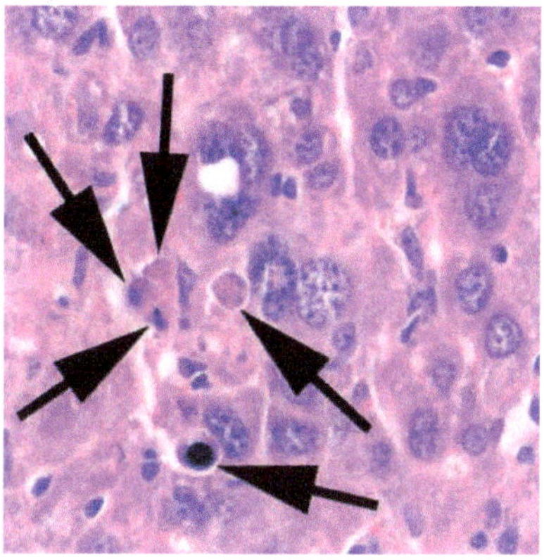

A section of mouse liver showing several apoptotic cells, indicated by arrows

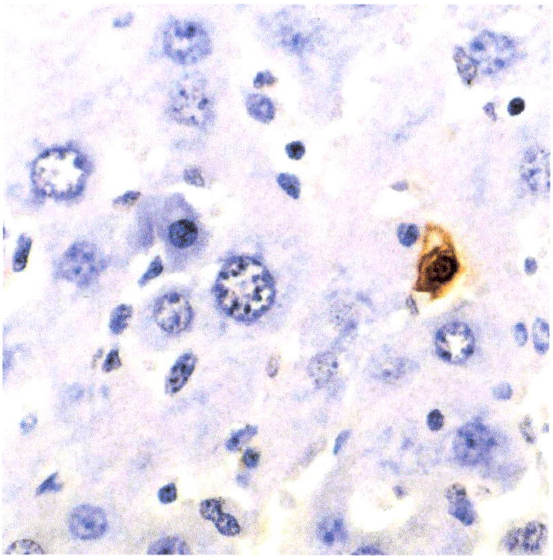

A section of mouse liver stained to show cells undergoing apoptosis (orange)

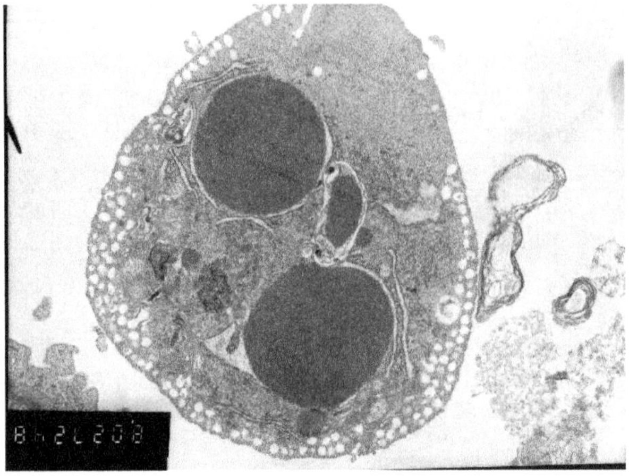

Neonatal cardiomyocytes ultrastructure after anoxia-reoxygenation.

Defective Pathways

The many different types of apoptotic pathways contain a multitude of different biochemical components, many of them not yet understood. As a pathway is more or less sequential in nature, removing or modifying one component leads to an effect in another. In a living organism, this can have disastrous effects, often in the form of disease or disorder. A discussion of every disease caused by modification of the various apoptotic pathways would be impractical, but the concept overlying each one is the same: The normal functioning of the pathway has been disrupted in such a way as to impair the ability of the cell to undergo normal apoptosis. This results in a cell that lives past its "use-by-date" and is able to replicate and pass on any faulty machinery to its progeny, increasing the likelihood of the cell's becoming cancerous or diseased.

A recently described example of this concept in action can be seen in the development of a lung cancer called NCI-H460. The *X-linked inhibitor of apoptosis protein* (XIAP) is overexpressed in cells of the H460 cell line. XIAPs bind to the processed form of caspase-9, and suppress the activity of apoptotic activator cytochrome c, therefore overexpression leads to a decrease in the amount of proapoptotic agonists. As a consequence, the balance of anti-apoptotic and proapoptotic effectors is upset in favour of the former, and the damaged cells continue to replicate despite being directed to die.

Dysregulation of p53

The tumor-suppressor protein p53 accumulates when DNA is damaged due to a chain of biochemical factors. Part of this pathway includes alpha-interferon and beta-interferon, which induce transcription of the *p53* gene, resulting in the increase of p53 protein level and enhancement of cancer cell-apoptosis. p53 prevents the cell from replicating by stopping the cell cycle at G1, or interphase, to give the cell time to repair, however it will induce apoptosis if damage is extensive and repair efforts fail. Any disruption to the regulation of the *p53* or interferon genes will result in impaired apoptosis and the possible formation of tumors.

Inhibition

Inhibition of apoptosis can result in a number of cancers, autoimmune diseases, inflammatory diseases, and viral infections. It was originally believed that the associated accumulation of cells was due to an increase in cellular proliferation, but it is now known that it is also due to a decrease in cell death. The most common of these diseases is cancer, the disease of excessive cellular proliferation, which is often characterized by an overexpression of IAP family members. As a result, the malignant cells experience an abnormal response to apoptosis induction: Cycle-regulating genes (such as p53, ras or c-myc) are mutated or inactivated in diseased cells, and further genes (such as bcl-2) also modify their expression in tumors.

HeLa cell

Apoptosis in HeLacells is inhibited by proteins produced by the cell; these inhibitory proteins target retinoblastoma tumor-suppressing proteins. These tumor-suppressing proteins regulate the cell cycle, but are rendered inactive when bound to an inhibitory protein. HPV E6 and E7 are inhibitory proteins expressed by the human papillomavirus, HPV being responsible for the formation of the cervical tumor from which HeLa cells are derived. HPV E6 causes p53, which regulates the cell cycle, to become inactive. HPV E7 binds to retinoblastoma tumor suppressing proteins and limits its ability to control cell division. These two inhibitory proteins are partially responsible for HeLa cells' immortality by inhibiting apoptosis to occur. CDV (Canine Distemper Virus) is able to induce apoptosis despite the presence of these inhibitory proteins. This is an important oncolytic property of CDV: this virus is capable of killing canine lymphoma cells. Oncoproteins E6 and E7 still leave p53 inactive, but they are not able to avoid the activation of caspases induced from the stress of viral infection. These oncolytic properties provided a promising link between CDV and lymphoma apoptosis, which can lead to development of alternative treatment methods for both canine lymphoma and human non-Hodgkin lymphoma. Defects in the cell cycle are thought to be responsible for the resistance to chemotherapy or radiation by certain tumor cells, so a virus that can induce apoptosis despite defects in the cell cycle is useful for cancer treatment.

Treatments

The main method of treatment for death from signaling-related diseases involves either increasing or decreasing the susceptibility of apoptosis in diseased cells, depending on whether the disease is caused by either the inhibition of or excess apoptosis. For instance, treatments aim to restore apoptosis to treat diseases with deficient cell death, and to increase the apoptotic threshold to treat diseases involved with excessive cell death. To stimulate apoptosis, one can increase the number of death receptor ligands (such as TNF or TRAIL), antagonize the anti-apoptotic Bcl-2 pathway, or introduce Smac mimetics to inhibit the inhibitor (IAPs). The addition of agents such as Herceptin, Iressa, or Gleevec works to stop cells from cycling and causes apoptosis activation by blocking growth and survival signaling further upstream. Finally, adding p53-MDM2 complexes displaces p53 and activates the p53 pathway, leading to cell cycle arrest and apoptosis. Many different methods can be used either to stimulate or to inhibit apoptosis in various places along the death signaling pathway.

Apoptosis is a multi-step, multi-pathway cell-death programme that is inherent in every cell of the body. In cancer, the apoptosis cell-division ratio is altered. Cancer treatment by chemotherapy and irradiation kills target cells primarily by inducing apoptosis.

Hyperactive Apoptosis

On the other hand, loss of control of cell death (resulting in excess apoptosis) can lead to neurodegenerative diseases, hematologic diseases, and tissue damage. The progression of HIV is directly linked to excess, unregulated apoptosis. In a healthy individual, the number of CD4+ lymphocytes is in balance with the cells generated by the bone marrow; however, in HIV-positive patients, this balance is lost due to an inability of the bone marrow to regenerate CD4+ cells. In the case of HIV, CD4+ lymphocytes die at an accelerated rate through uncontrolled apoptosis, when stimulated.

Treatments

Treatments aiming to inhibit works to block specific caspases. Finally, the Akt protein kinase promotes cell survival through two pathways. Akt phosphorylates and inhibits Bas (a Bcl-2 family member), causing Bas to interact with the 14-3-3 scaffold, resulting in Bcl dissociation and thus cell survival. Akt also activates IKKα, which leads to NF-κB activation and cell survival. Active NF-κB induces the expression of anti-apoptotic genes such as Bcl-2, resulting in inhibition of apoptosis. NF-κB has been found to play both an antiapoptotic role and a proapoptotic role depending on the stimuli utilized and the cell type.

HIV Progression

The progression of the human immunodeficiency virus infection into AIDS is due primarily to the depletion of CD4+ T-helper lymphocytes in a manner that is too rapid for the body's bone marrow to replenish the cells, leading to a compromised immune system. One of the mechanisms by which T-helper cells are depleted is apoptosis, which results from a series of biochemical pathways:

- HIV enzymes deactivate anti-apoptotic *Bcl-2*. This does not directly cause cell death but primes the cell for apoptosis should the appropriate signal be received. In parallel, these enzymes activate proapoptotic *procaspase-8*, which does directly activate the mitochondrial events of apoptosis.
- HIV may increase the level of cellular proteins that prompt Fas-mediated apoptosis.
- HIV proteins decrease the amount of CD4 glycoprotein marker present on the cell membrane.

- Released viral particles and proteins present in extracellular fluid are able to induce apoptosis in nearby bystander T helper cells.
- HIV decreases the production of molecules involved in marking the cell for apoptosis, giving the virus time to replicate and continue releasing apoptotic agents and virions into the surrounding tissue.
- The infected CD4+ cell may also receive the death signal from a cytotoxic T cell.

Cells may also die as direct consequences of viral infections. HIV-1 expression induces tubular cell G2/M arrest and apoptosis. The progression from HIV to AIDS is not immediate or even necessarily rapid; HIV's cytotoxic activity toward CD4+ lymphocytes is classified as AIDS once a given patient's CD4+ cell count falls below 200.

Viral Infection

Viral induction of apoptosis occurs when one or several cells of a living organism are infected with a virus, leading to cell death. Cell death in organisms is necessary for the normal development of cells and the cell cycle maturation. It is also important in maintaining the regular functions and activities of cells.

Viruses can trigger apoptosis of infected cells via a range of mechanisms including:

- Receptor binding
- Activation of protein kinase R (PKR)
- Interaction with p53
- Expression of viral proteins coupled to MHC proteins on the surface of the infected cell, allowing recognition by cells of the immune system (such as Natural Killer and cytotoxic T cells) that then induce the infected cell to undergo apoptosis.

Canine distemper virus (CDV) is known to cause apoptosis in central nervous system and lymphoid tissue of infected dogs in vivo and in vitro.Apoptosis caused by CDV is typically induced via the extrinsic pathway, which activates caspases that disrupt cellular function and eventually leads to the cells death. In normal cells, CDV activates caspase-8 first, which works as the initiator protein followed by the executioner protein caspase-3. However, apoptosis induced by CDV in HeLa cells does not involve the initiator protein caspase-8. HeLa cell apoptosis caused by CDV follows a different mechanism than that in vero cell lines. This change in the caspase cascade suggests CDV induces apoptosis via the intrinsic pathway, excluding the need for the initiator caspase-8. The executioner protein is instead activated by the internal stimuli caused by viral infection not a caspase cascade.

The Oropouche virus (OROV) is found in the family *Bunyaviridae*. The study of apoptosis brought on by *Bunyaviridae* was initiated in 1996, when it was observed that apoptosis was induced by the La Crosse virus into the kidney cells of baby hamsters and into the brains of baby mice.

OROV is a disease that is transmitted between humans by the biting midge (*Culicoides paraensis*). It is referred to as a zoonotic arbovirus and causes febrile illness, characterized by the onset of a sudden fever known as Oropouche fever.

The Oropouche virus also causes disruption in cultured cells – cells that are cultivated in distinct and specific conditions. An example of this can be seen in HeLa cells, whereby the cells begin to degenerate shortly after they are infected.

With the use of gel electrophoresis, it can be observed that OROV causes DNA fragmentation in HeLa cells. It can be interpreted by counting, measuring, and analyzing the cells of the Sub/G1 cell population. When HeLA cells are infected with OROV, the cytochrome C is released from the membrane of the mitochondria, into the cytosol of the cells. This type of interaction shows that apoptosis is activated via an intrinsic pathway.

In order for apoptosis to occur within OROV, viral uncoating, viral internalization, along with the replication of cells is necessary. Apoptosis in some viruses is activated by extracellular stimuli. However, studies have demonstrated that the OROV infection causes apoptosis to be activated through intracellular stimuli and involves the mitochondria.

Many viruses encode proteins that can inhibit apoptosis. Several viruses encode viral homologs of Bcl-2. These homologs can inhibit proapoptotic proteins such as BAX and BAK, which are essential for the activation of apoptosis. Examples of viral Bcl-2 proteins include the Epstein-Barr virus BHRF1 protein and the adenovirus E1B 19K protein. Some viruses express caspase inhibitors that inhibit caspase activity and an example is the CrmA protein of cowpox viruses. Whilst a number of viruses can block the effects of TNF and Fas. For example, the M-T2 protein of myxoma viruses can bind TNF preventing it from binding the TNF receptor and inducing a response. Furthermore, many viruses express p53 inhibitors that can bind p53 and inhibit its transcriptional transactivation activity. As a consequence, p53 cannot induce apoptosis, since it cannot induce the expression of proapoptotic proteins. The adenovirus E1B-55K protein and the hepatitis B virus HBx protein are examples of viral proteins that can perform such a function.

Viruses can remain intact from apoptosis in particular in the latter stages of infection. They can be exported in the *apoptotic bodies* that pinch off from the surface of the dying cell, and the fact that they are engulfed by phagocytes prevents the initiation of a host response. This favours the spread of the virus.

Plants

Programmed cell death in plants has a number of molecular similarities to that of animal apoptosis, but it also has differences, notable ones being the presence of a cell wall and the lack of an immune system that removes the pieces of the dead cell. Instead of an immune response, the dying cell synthesizes substances to break itself down and places them in a vacuole that ruptures as the cell dies. Whether this whole process resembles animal apoptosis closely enough to warrant using the name *apoptosis* (as opposed to the more general *programmed cell death*) is unclear.

Caspase-Independent Apoptosis

The characterization of the caspases allowed the development of caspase inhibitors, which can be used to determine whether a cellular process involves active caspases. Using these inhibitors it was discovered that cells can die while displaying a morphology similar to apoptosis without caspase activation. Later studies linked this phenomenon to the release of AIF (apoptosis-inducing factor) from the mitochondria and its translocation into the nucleus mediated by its NLS (nuclear localization signal). Inside the mitochondria, AIF is anchored to the inner membrane. In order to be released, the protein is cleaved by a calcium-dependent calpain protease.

Apoptosis Protein Subcellular Location Prediction

In 2003, a method was developed for predicting subcellular location of apoptosis proteins.[82] Subsequent to this, various modes of Chou's pseudo amino acid composition were developed for improving the quality of predicting subcellular localization of apoptosis proteins based on their sequence information alone.

REFERENCES

1. Freeman, Scott (2011). Biological Science (6th ed.). Hoboken, NY: Pearson. p. 210.
2. Letunic, I; Bork, P (2006). Interactive Tree of Life. Retrieved 23 July 2011.
3. Bernstein H, Bernstein C, Michod RE (2011). "Meiosis as an evolutionary adaptation for DNA repair." In "DNA Repair", Intech Publ (Inna Kruman, editor), Chapter 19: 357-382 DOI: 10.5772/1751 ISBN 978-953-307-697-3 Available online from: http://www.intechopen.com/books/dna-repair/meiosis-as-an-evolutionary-adaptation-for-dna-repair
4. Bernstein H, Bernstein C (2010). Evolutionary origin of recombination during meiosis. BioScience. 60 (7): 498–505. doi:10.1525/bio.2010.60.7.5.
5. LODÉ T (2011). Sex is not a solution for reproduction: the libertine bubble theory. BioEssays. 33 (6): 419–422. doi:10.1002/bies.201000125. PMID 21472739.
6. Hassold, Terry; Hunt, Patricia (1 April 2001). To err (meiotically) is human: the genesis of human aneuploidy. Nature Reviews Genetics. 2 (4): 280–291. doi:10.1038/35066065. PMID 11283700.
7. J.B. Farmer and J.E.S. Moore, Quarterly Journal of Microscopic Science 48:489 (1905) as quoted in the Oxford English Dictionary, Third Edition, June 2001, s.v.
8. Zhou, A.; Pawlowski, W.P. (August 2014). Regulation of meiotic gene expression in plants. Frontiers in Plant Science. 5: Article 413. doi:10.3389/fpls.2014.00413. PMID 25202317.
9. Brar GA, Yassour M, Friedman N, Regev A, Ingolia NT, Weissman JS (February 2012). High-resolution view of the yeast meiotic program revealed by ribosome profiling. Science. 335 (6068): 552–7. doi:10.1126/science.1215110. PMC 3414261. PMID 22194413.
10. Freeman 2005, pp. 244–45
11. Freeman 2005, pp. 249–250
12. Snustad, DP; Simmons, MJ (December 2008). Principles of Genetics (5th ed.).

Wiley. ISBN 978-0-470-38825-9.

13. Krebs, JE; Goldstein, ES; Kilpatrick, ST (November 2009). Lewins Genes X (10th ed.). Jones & Barlett Learning. ISBN 978-0-7637-6632-0.
14. Raven, Peter H.; Johnson, George B.; Mason, Kenneth A.; Losos, Jonathan & Singer, Susan. Biology, 8th ed. McGraw-Hill 2007.
15. Petronczki M, Siomos MF, Nasmyth K (February 2003). Un ménage à quatre: the molecular biology of chromosome segregation in meiosis. Cell. 112 (4): 423–40. doi:10.1016/S0092-8674(03)00083-7. PMID 12600308.
16. Retallack GJ, Krull ES, Thackeray GD, Parkinson D (2013-09-01). Problematic urn-shaped fossils from a Paleoproterozoic (2.2 Ga) paleosol in South Africa. Precambrian Research. 235: 71–87. doi:10.1016/j.precamres.2013.05.015.
17. Hassold, T.; Jacobs, P.; Kline, J.; Stein, Z.; Warburton, D. (July 1980). Effect of maternal age on autosomal trisomies. Annals of Human Genetics. 44 (1): 29–36. doi:10.1111/j.1469-1809.1980.tb00943.x. PMID 7198887.
18. Tsutsumi, M.; Fujiwara, R.; Nishizawa, H.; Ito, M.; Kogo, H.; Inagaki, H.; Ohye, T.; Kato, T.; Fujii, T.; Kurahashi, H. (May 2014). Age-related decrease of meiotic cohesins in human oocytes. PLOS ONE. 9 (5): Article e96710. doi:10.1371/journal.pone.0096710. PMC 4013030. PMID 24806359.
19. Bidlack, James E. (2011). Introductory Plant Biology. New York, NY: McGraw HIll. pp. 214–29.
20. Brunet, S.; Verlhac, M. H. (2010). Positioning to get out of meiosis: The asymmetry of division. Human Reproduction Update. 17 (1): 68–75. doi:10.1093/humupd/dmq044. PMID 20833637.
21. Rosenbusch B (November 2006). The contradictory information on the distribution of non-disjunction and pre-division in female gametes. Hum. Reprod. 21 (11): 2739–42. doi:10.1093/humrep/del122. PMID 16982661.
22. Lin Y, Gill ME, Koubova J, Page DC (December 2008). Germ cell-intrinsic and -extrinsic factors govern meiotic initiation in mouse embryos. Science. 322 (5908): 1685–7. doi:10.1126/science.1166340. PMID 19074348.
23. Suzuki A, Saga Y (February 2008). Nanos2 suppresses meiosis and promotes male germ cell differentiation. Genes Dev. 22 (4): 430–5. doi:10.1101/gad.1612708. PMC 2238665. PMID 18281459.
24. How Cells Divide. PBS. Public Broadcasting Service. Retrieved 6 December 2012.
25. Heywood, P.; Magee, P.T. (1976). Meiosis in protists. Some structural and physiological aspects of meiosis in algae, fungi, and protozoa. Bacteriological Reviews. 40 (1): 190–240.
26. Raikov, I. B. (1995). Meiosis in protists: recent advances and persisting problems. Europ J Protistol. 31: 1–7. doi:10.1016/s0932-4739(11)80349-4.
27. "Quantitative phase contrast microscopy — label-free live cell imaging and quantification". Phase Holographic Imaging AB.
28. Green, Douglas (2011). Means to an End: Apoptosis and other Cell Death

Mechanisms. Cold Spring Harbor, NY: Cold Spring Harbor Laboratory Press. ISBN 978-0-87969-888-1.

29. Alberts, p. 2.

30. Karam, Jose A. (2009). Apoptosis in Carcinogenesis and Chemotherapy. Netherlands: Springer. ISBN 978-1-4020-9597-9.

31. Alberts, Bruce; Johnson, Alexander; Lewis, Julian; Raff, Martin; Roberts, Keith; Walter, Peter (2008). "Chapter 18 Apoptosis: Programmed Cell Death Eliminates Unwanted Cells". Molecular Biology of the Cell (textbook) (5th ed.). Garland Science. p. 1115. ISBN 978-0-8153-4105-5.

32. Kerr, JF. (1965). "A histochemical study of hypertrophy and ischaemic injury of rat liver with special reference to changes in lysosomes". Journal of Pathology and Bacteriology. 90 (2): 419–35. doi:10.1002/path.1700900210.

33. Agency for Science, Technology and Research. "Prof Andrew H. Wyllie – Lecture Abstract". Archived from the original on 2007-11-13. Retrieved 2007-03-30.

34. Kerr JF, Wyllie AH, Currie AR (August 1972). "Apoptosis: a basic biological phenomenon with wide-ranging implications in tissue kinetics". Br. J. Cancer. 26 (4): 239–57. doi:10.1038/bjc.1972.33. PMC 2008650. PMID 4561027.

35. O'Rourke MG, Ellem KA (2000). "John Kerr and apoptosis". Med. J. Aust. 173 (11–12): 616–7. PMID 11379508.

36. Alberts, p. 1021.

37. American Heritage Dictionary Archived June 30, 2008, at the Wayback Machine.

38. Apoptosis Interest Group (1999). "About apoptosis". Archivedfrom the original on 28 December 2006. Retrieved 2006-12-15.

39. Webster.com dictionary entry

40. Alberts, p. 1029.

41. Böhm I, Schild H (2003). "Apoptosis: the complex scenario for a silent cell death". Mol Imaging Biol. 5 (1): 2–14. doi:10.1016/S1536-1632(03)00024-6. PMID 14499155.

42. Cotran, Ramzi, S.; Kumar, Collins (1998). Robbins Pathologic Basis of Disease. Philadelphia: W.B Saunders Company. ISBN 0-7216-7335-X.

43. Mattson MP, Chan SL (2003). "Calcium orchestrates apoptosis". Nature Cell Biology. 5 (12): 1041–1043. doi:10.1038/ncb1203-1041. PMID 14647298.

44. Uguz AC, Naziroglu M, Espino J, Bejarano I, González D, Rodríguez AB, Pariente JA (November 2009). "Selenium Modulates Oxidative Stress-Induced Cell Apoptosis in Human Myeloid HL-60 Cells Through Regulationof Calcium Release and Caspase-3 and -9 Activities". J Membrane Biol 232: 15-23. doi:[10.1007/s00232-009-9212-2].

45. Chiarugi A, Moskowitz MA (2002). "PARP-1—a perpetrator of apoptotic cell death?. Science. 297 (5579): 259–63. doi:10.1126/science.1074592. PMID 12114611.

46. Bejarano I, Espino J, González-Flores D, Casado JG, Redondo PC, Rosado

JA, Barriga C, Pariente JA, Rodríguez AB (2009). "Role of Calcium Signals on Hydrogen Peroxide-Induced Apoptosis in Human Myeloid HL-60 Cells". International Journal of Biomedical Science. 5 (3): 246–256. PMC 3614781. PMID 23675144.

47. D. González, I. Bejarano, C. Barriga, A.B. Rodríguez, J.A. Pariente (2010). "Oxidative Stress-Induced Caspases are Regulated in Human Myeloid HL-60 Cells by Calcium Signal. Current Signal Transduction Therapy 5: 181-186. do i:[10.2174/157436210791112172]
48. Mohan S, Bustamam A, Abdelwahab SI, Al-Zubairi AS, Aspollah M, Abdullah R, Elhassan MM, Ibrahim MY, Syam S: Typhonium flagelliforme induces apoptosis in CEMss cells via activation of caspase-9, PARP cleavage and cytochrome c release: Its activation coupled with G0/G1 phase cell cycle arrest. Journal of Ethnopharmacology 2010
49. Brüne B (August 2003). "Nitric oxide: NO apoptosis or turning it ON?. Cell Death Differ. 10 (8): 864–9. doi:10.1038/sj.cdd.4401261. PMID 12867993.
50. Brune B, et al. (1999). "Nitric oxide (NO): an effector of apoptosis". Cell Death and Differentiation. 6 (10): 969–975. doi:10.1038/sj.cdd.4400582.
51. Fesik SW, Shi Y (2001). "Controlling the caspases". Science. 294 (5546): 1477–8. doi:10.1126/science.1062236. PMID 11711663.
52. Laurent M. Dejean; Sonia Martinez-Caballero; Kathleen W. Kinnally (2006). "Is MAC the knife that cuts cytochrome c from mitochondria during apoptosis?. Cell Death and Differentiation. 13 (8): 1387–5. doi:10.1038/sj.cdd.4401949. PMID 16676005.
53. Dejean LM, Martinez-Caballero S, Manon S, Kinnally KW (February 2006). "Regulation of the mitochondrial apoptosis-induced channel, MAC, by BCL-2 family proteins". Biochim. Biophys. Acta. 1762 (2): 191–201. doi:10.1016/j.bbadis.2005.07.002. PMID 16055309.
54. Lodish, Harvey; et al. (2004). Molecular Cell Biology. New York: W.H. Freedman and Company. ISBN 0-7167-4366-3.
55. Wajant H (2002). The Fas signaling pathway: more than a paradigm. Science. 296 (5573): 1635–6. doi:10.1126/science.1071553. PMID 12040174.
56. Chen G, Goeddel DV (2002). "TNF-R1 signaling: a beautiful pathway". Science. 296 (5573): 1634–5. doi:10.1126/science.1071924. PMID 12040173.
57. Goeddel, DV. "Connection Map for Tumor Necrosis Factor Pathway". Science STKE. 2007: tw132. doi:10.1126/stke.3822007tw132.
58. Chen W, Li N, Chen T, Han Y, Li C, Wang Y, et al. (2005). "The lysosome-associated apoptosis-inducing protein containing the pleckstrin homology (PH) and FYVE domains (LAPF), representative of a novel family of PH and FYVE domain-containing proteins, induces caspase-independent apoptosis via the lysosomal-mitochondrial pathway.. Journal of Biological Chemistry. 280 (49): 40985–40995. doi:10.1074/jbc.M502190200. PMID 16188880.
59. Wajant, H. "Connection Map for Fas Signaling Pathway". Science STKE. 2007:

trl. doi:10.1126/stke.3802007trl.

60. Murphy KM, Ranganathan V, Farnsworth ML, Kavallaris M, Lock RB (January 2000). “Bcl-2 inhibits Bax translocation from cytosol to mitochondria during drug-induced apoptosis of human tumor cells”. Cell Death Differ. 7 (1): 102–11. doi:10.1038/sj.cdd.4400597. PMID 10713725.

61. Susin SA, Lorenzo HK, Zamzami N (February 1999). “Molecular characterization of mitochondrial apoptosis-inducing factor”. Nature. 397 (6718): 441–6. doi:10.1038/17135. PMID 9989411.

62. Jewhurst K, Levin M, McLaughlin KA (2014). “Optogenetic Control of Apoptosis in Targeted Tissues of Xenopus laevis Embryos”. J Cell Death. 7: 25–31. doi:10.4137/JCD.S18368. PMC 4213186. PMID 25374461.

63. Venturi, Sebastiano (2011). “Evolutionary Significance of Iodine”. Current Chemical Biology-. 5 (3): 155–162. doi:10.2174/187231311796765012. ISSN 1872-3136.

64. Venturi, Sebastiano (2014). “Iodine, PUFAs and Iodolipids in Health and Disease: An Evolutionary Perspective”. Human Evolution-. 29 (1-3): 185–205. ISSN 0393-9375.

65. Tamura K, Takayama S, Ishii T, Mawaribuchi S, Takamatsu N, Ito M (2015). “Apoptosis and differentiation of Xenopus tail-derived myoblasts by thyroid hormone.. J Mol Endocrinol. 54 (3): 185–92. doi:10.1530/JME-14-0327. PMID 25791374.

66. Thomas, M. P.; Liu, X; Whangbo, J; McCrossan, G; Sanborn, K. B.; Basar, E; Walch, M; Lieberman, J (2015). “Apoptosis Triggers Specific, Rapid, and Global mRNA Decay with 3’ Uridylated Intermediates Degraded by DIS3L2”. Cell Reports. 11: 1079–89. doi:10.1016/j.celrep.2015.04.026. PMID 25959823.

67. Böhm I (2003). “Disruption of the cytoskeleton after apoptosis induction by autoantibodies”. Autoimmunity. 36: 183–189. doi:10.1080/0891693031000105617.

68. Susin, S; Daugas, E; Ravagnan, L; Samejima, K; Zamzami, N; Loeffler, M; Costantini, P; Ferri, KF; et al. (2000). “Two Distinct Pathways Leading to Nuclear Apoptosis”. Journal of Experimental Medicine. 192 (4): 571–80. doi:10.1084/jem.192.4.571. PMC 2193229. PMID 10952727.

69. Madeleine Kihlmark; Imreh, G; Hallberg, E (15 October 2001). “Sequential degradation of proteins from the nuclear envelope during apoptosis”. Journal of Cell Science. 114 (20): 3643–53. PMID 11707516.

70. Nagata S (April 2000). “Apoptotic DNA fragmentation”. Exp. Cell Res. 256 (1): 12–8. doi:10.1006/excr.2000.4834. PMID 10739646.

71. M Iwata; D Myerson; B Torok-Storb; RA Zager (1996). “An evaluation of renal tubular DNA laddering in response to oxygen deprivation and oxidant injury”. Retrieved 2006-04-17.

72. Vandivier RW, Henson PM, Douglas IS (June 2006). “Burying the dead: the impact of failed apoptotic cell removal (efferocytosis) on chronic inflammatory lung disease”. Chest. 129 (6): 1673–82. doi:10.1378/chest.129.6.1673. PMID

16778289.

73. Li MO; Sarkisian, MR; Mehal, WZ; Rakic, P; Flavell, RA (2003). "Phosphatidylserine receptor is required for clearance of apoptotic cells". Science. 302 (5650): 1560–3. doi:10.1126/science.1087621. PMID 14645847.

74. Wang X; Wu, YC; Fadok, VA; Lee, MC; Gengyo-Ando, K; Cheng, LC; Ledwich, D; Hsu, PK; et al. (2003). "Cell corpse engulfment mediated by C. elegans phosphatidylserine receptor through CED-5 and CED-12". Science. 302 (5650): 1563–1566. doi:10.1126/science.1087641. PMID 14645848.

75. Savill J, Gregory C, Haslett C. (2003). "Eat me or die". Science. 302 (5650): 1516–7. doi:10.1126/science.1092533. PMID 14645835.

76. Krysko DV; Vandenabeele P. Phagocytosis of dying cells: from molecular mechanisms to human diseases. ISBN 978-1-4020-9292-3.

77. Lozano GM, Bejarano I, Espino J, González D, Ortiz A, García JF, Rodríguez AB, Pariente JA (2009). "Density gradient capacitation is the most suitable method to improve fertilization and to reduce DNA fragmentation positive spermatozoa of infertile men". Anatolian Journal of Obstetrics & Gynecology. 3 (1): 1–7.

78. Krysko DV, Vanden Berghe T, Parthoens E, D'Herde K, Vandenabeele P (2008). "Methods for distinguishing apoptotic from necrotic cells and measuring their clearance.. Methods Enzymol. 442: 307–41. doi:10.1016/S0076-6879(08)01416-X. PMID 18662577.

79. Krysko DV, Vanden Berghe T, D'Herde K, Vandenabeele P (2008). "Apoptosis and necrosis: detection, discrimination and phagocytosis.. Methods. 44 (3): 205–21. doi:10.1016/j.ymeth.2007.12.001. PMID 18314051.

80. Vanden Berghe T, Grootjans S, Goossens V, Dondelinger Y, Krysko DV, Takahashi N, Vandenabeele P (2013). "Determination of apoptotic and necrotic cell death in vitro and in vivo.. Methods. 61 (2): 117–29. doi:10.1016/j.ymeth.2013.02.011. PMID 23473780.

81. Thompson, CB (1995). "Apoptosis in the pathogenesis and treatment of disease". Science. 267 (5203): 1456–62. doi:10.1126/science.7878464. PMID 7878464.

82. Yang L, Mashima T, Sato S (February 2003). "Predominant suppression of apoptosome by inhibitor of apoptosis protein in non-small cell lung cancer H460 cells: therapeutic effect of a novel polyarginine-conjugated Smac peptide". Cancer Res. 63 (4): 831–7. PMID 12591734.

83. Takaoka A; Hayakawa, S; Yanai, H; Stoiber, D; Negishi, H; Kikuchi, H; Sasaki, S; Imai, K; et al. (2003). "Integration of interferon-alpha/beta signalling to p53 responses in tumour suppression and antiviral defence". Nature. 424 (6948): 516–23. doi:10.1038/nature01850. PMID 12872134.

84. Del Puerto HL, Martins AS, Milsted A, et al. (2011). Canine distemper virus induces apoptosis in cervical tumor derived cell lines. Virol. J. 8 (1): 334. doi:10.1186/1743-422X-8-334. PMC 3141686. PMID 21718481.

85. Liu HC, Chen GG, Vlantis AC, Tse GM, Chan AT, van Hasselt CA (March 2008). "Inhibition of apoptosis in human laryngeal cancer cells by E6 and E7

oncoproteins of human papillomavirus 16". J. Cell. Biochem. 103 (4): 1125–43. doi:10.1002/jcb.21490. PMID 17668439.

86. Niu XY, Peng ZL, Duan WQ, Wang H, Wang P (2006). Inhibition of HPV 16 E6 oncogene expression by RNA interference in vitro and in vivo. Int. J. Gynecol. Cancer. 16 (2): 743–51. doi:10.1111/j.1525-1438.2006.00384.x. PMID 16681755.
87. Liu, Y; McKalip, A; Herman, B (2000). Human Papillomavirus Type 16 E6 and HPV-16 E6/E7 Sensitize Human Keratinocytes to Apoptosis Induced by Chemotherapeutic Agents: Roles of p53 and Caspase Activation. Journal of Cellular Biochemistry. 78: 334–349. doi:10.1002/(sici)1097-4644(20000801)78:2<334::aid-jcb15>3.3.co;2-6.
88. Boehm, I. (2006). "Apoptosis in physiological and pathological skin: Implications for therapy". Current Molecular Medicine. 6 (4): 375–394. doi:10.2174/156652406777435390. PMID 16900661.
89. Farhana, Lulu. "Apoptosis Induction by a Novel Retinoid-Related Molecule Requires Nuclear Factor- κB Activation" (PDF). American Association of Cancer Research (AACR). Retrieved 2012-01-24.
90. Judie B. Alimonti; T. Blake Ball; Keith R. Fowke (2003). "Mechanisms of CD4+ T lymphocyte cell death in human immunodeficiency virus infection and AIDS". J. Gen. Virology. 84 (7): 1649–61. doi:10.1099/vir.0.19110-0. PMID 12810858.
91. Vashistha H, Husain M, Kumar D, Yadav A, Arora S, Singhal PC. (2008) Ren Fail. 2008;30(6):655-64.
92. Indiana University Health. "AIDS Defining Criteria | Riley". IU Health. Retrieved 2013-01-20.
93. Indran IR, Tufo G, Pervaiz S, Brenner C (June 2011). Recent advances in apoptosis, mitochondria and drug resistance in cancer cells. Biochim. Biophys. Acta. 1807 (6): 735–45. doi:10.1016/j.bbabio.2011.03.010. PMID 21453675.
94. Everett, H.; McFadden, G. (1999). "Apoptosis: an innate immune response to virus infection". Trends Microbiol. 7 (4): 160–5. doi:10.1016/S0966-842X(99)01487-0. PMID 10217831.
95. Nishi T, Tsukiyama-Kohara K, Togashi K, Kohriyama N, Kai C (November 2004). "Involvement of apoptosis in syncytial cell death induced by canine distemper virus". Comp. Immunol. Microbiol. Infect. Dis. 27 (6): 445–55. doi:10.1016/j.cimid.2004.01.007. PMID 15325517.
96. Acrani GO, Gomes R, Proença-Módena JL, et al. (April 2010). Apoptosis induced by Oropouche virus infection in HeLa cells is dependent on virus protein expression. Virus Res. 149 (1): 56–63. doi:10.1016/j.virusres.2009.12.013. PMID 20080135.
97. Azevedo RS, Nunes MR, Chiang JO, et al. (June 2007). "Reemergence of Oropouche fever, northern Brazil". Emerging Infect. Dis. 13 (6): 912–5. doi:10.3201/eid1306.061114. PMC 2792853. PMID 17553235.
98. Santos RI, Rodrigues AH, Silva ML, et al. (December 2008). "Oropouche virus entry into HeLa cells involves clathrin and requires endosomal acidification". Virus

Res. 138 (1–2): 139–43. doi:10.1016/j.virusres.2008.08.016. PMID 18840482.

99. Teodoro, J.G. Branton, P.E. (1997). "Regulation of apoptosis by viral gene products". J. Virol. 71 (3): 1739–46. PMC 191242. PMID 9032302.

100. Polster, B.M. Pevsner, J. and Hardwick, J.M. (2004). "Viral Bcl-2 homologs and their role in virus replication and associated diseases". Biochim. Biophys. Acta. 1644 (2–3): 211–27. doi:10.1016/j.bbamcr.2003.11.001. PMID 14996505.

101. Hay, S.; Kannourakis, G. (2002). A time to kill: viral manipulation of the cell death program. J. Gen. Virol. 83 (Pt 7): 1547–64. doi:10.1099/0022-1317-83-7-1547. PMID 12075073.

102. Wang, X.W. Gibson, M.K. Vermeulen, W. Yeh, H. Forrester, K. Sturzbecher, H.W. Hoeijmakers, J.H. and Harris, C.C. (1995). "Abrogation of p53-induced Apoptosis by the Hepatitis B Virus X Gene". Cancer Res. 55 (24): 6012–6. PMID 8521383.

103. Cyrelys Collazo; Osmani Chacón; Orlando Borrás (2006). "Programmed cell death in plants resembles apoptosis of animals"(PDF). Biotecnología Aplicada. 23: 1–10.

104. Xiang J. et al., BAX-induced cell death may not require interleukin 1β-converting enzyme-likeproteases, 1996, cell biology

105. Zhou, G. P. & Doctor, K. (2003). Subcellular location prediction of apoptosis proteins. PROTEINS: Structure, Function, and Genetics 50, 44-48.

106. Chen Y. L.; Li Q. Z. (2007). "Prediction of apoptosis protein subcellular location using improved hybrid approach and pseudo amino acid composition". Journal of Theoretical Biology. 248 (2): 377–381. doi:10.1016/j.jtbi.2007.05.019. PMID 17572445.

107. Ding, Y. S. & Zhang, T. L. (2008). Using Chou's pseudo amino acid composition to predict subcellular localization of apoptosis proteins: an approach with immune genetic algorithm-based ensemble classifier. Pattern Recognition Letters 29, 1887-1892.

108. Jiang, X., Wei, R., Zhang, T. L. & Gu, Q. (2008). Using the concept of Chous pseudo amino acid composition to predict apoptosis proteins subcellular location: an approach by approximate entropy. Protein & Peptide Letters 15, 392-396.

109. Lin H, Wang H, Ding H, Chen YL, Li QZ (2009). "Prediction of Subcellular Localization of Apoptosis Protein Using Chou's Pseudo Amino Acid Composition". Acta Biotheoretica. 57: 321–330. doi:10.1007/s10441-008-9067-4. PMID 19169652.

110. "Atromentin-Induced Apoptosis in Human Leukemia U937 Cells". Kim Jin Hee and Choong Hwan Lee, Journal of Microbiology and Biotechnology, 2009, vol. 19, no. 9, pages 946–950, INIST:21945937

111. Morgan D. (2006), The Cell Cycle: Principles of Control, OUP/New Science Press

112. Santos, S.D.M.; Ferrell, J.E. (2008), On the cell cycle and its switches, Nature, 454 (7202): 288, doi:10.1038/454288a, PMC 2727670, PMID 18633407

113. Skotheim, J.M.; Di Talia, S.; Siggia, E.D.; Cross, F.R. (2008), Positive feedback

of G1 cyclins ensures coherent cell cycle entry, Nature, 454 (7202): 291, doi:10.1038/nature07118, PMC 2606905, PMID 18633409

114. Pomerening J. R.; Sontag E. D.; et al. (2003). Building a cell cycle oscillator: hysteresis and bistability in the activation of Cdc2. Nat Cell Biol. 5 (4): 346–351. doi:10.1038/ncb954. PMID 12629549.

115. Holt L. J.; Krutchinsky A. N.; et al. (2008). Positive feedback sharpens the anaphase switch. Nature. 454 (7202): 353–357. doi:10.1038/nature07050. PMC 2636747. PMID 18552837.

116. Ferrell, J.E. (2008), "Feedback regulation of opposing enzymes generates robust, all-or-none bistable responses" (PDF), Current Biology, 18 (6): 244–245, doi:10.1016/j.cub.2008.02.035, PMC 2832910, PMID 18364225, retrieved 2009-12-11

117. Angeli, D.; Ferrell, J.E.; Sontag, E.D. (2004), "Detection of multistability, bifurcations and hysteresis in a large class of biological positive-feedback systems", Proceedings of the National Academy of Sciences, 101 (7): 1822, doi:10.1073/pnas.0308265100, retrieved 2009-12-11

118. Pfeuty B.; Kaneko K. (2009). "The combination of positive and negative feedback loops confers exquisite flexibility to biochemical switches". Phys. Biol. 046013: 1–11.

119. Kim SY; Ferrell JE (2007). Substrate competition as a source of ultrasensitivity in the activation of Wee1. Cell. 128 (6): 1133–45. doi:10.1016/j.cell.2007.01.039. PMID 17382882.

120. Strogatz S.H. (1994), Nonlinear Dynamics and Chaos, Perseus Books Publishing

121. Stuart, D.; Wittenberg, C. (1995), "CLN3, not positive feedback, determines the timing of CLN2 transcription in cycling cells." (PDF), Genes & Development, 9 (22): 2780, doi:10.1101/gad.9.22.2780, PMID 7590253, retrieved 2009-12-11

122. Harper JW (March 2002). "A phosphorylation-driven ubiquitination switch for cell-cycle control". Trends Cell Biol. 12 (3): 104–7. doi:10.1016/S0962-8924(01)02238-3. PMID 11859016.

123. Novak, B.; Tyson, J.J. (1993), "Numerical analysis of a comprehensive model of M-phase control in Xenopus oocyte extracts and intact embryos"(PDF), Journal of Cell Science, 106 (4): 1153, retrieved 2009-12-11

124. Sha, W.; Moore, J.; Chen, K.; Lassaletta, A.D.; Yi, C.S.; Tyson, J.J.; Sible, J.C. (2003), "Hysteresis drives cell-cycle transitions in Xenopus laevis egg extracts", Proceedings of the National Academy of Sciences, 100 (3): 975, doi:10.1073/pnas.0235349100, PMC 298711, PMID 12509509, retrieved 2009-12-11

125. Cooper, G. (2000), "The Cell: A Molecular Approach.", retrieved 2010-11-21

126. Uhlmann F.; Lottspeich F.; Nasmyth K. (1999). "Sister-chromatid separation at anaphase onset is promoted by cleavage of the cohesion subunit Scc1". Nature. 400 (6739): 37–42. doi:10.1038/21831. PMID 10403247.

127. Erich A. Nigg (2005). "Cyclin-dependent protein kinases: key regulators of the eukaryotic cell cycle". BioEssays. 17 (6): 471–480. doi:10.1002/bies.950170603.

PMID 7575488.

128. Mitosis#Cytokinesis
129. Sandra Lo´pez-Avile´s; Orsolya Kapuy; Bela Novak; Frank Uhlmann (2009). Irreversibility of mitotic exit is the consequence of systems-level feedback. Nature Letters. 459: 592–595. doi:10.1038/nature07984.
130. Randall W. King; Raymond J. Deshaies; Jan-Michael Peters; Marc W. Kirschner (1996). How proteolysis drives the cell cycle. Science. 274 (5293): 1652–1659. doi:10.1126/science.274.5293.1652. PMID 8939846.
131. Waizenegger; JF. Giménez-Abián; D. Wernic; JM. Peters (2002). Regulation of Human Separase by Securin Binding and Autocleavage. Current Biology. 12 (16): 1368–1378. doi:10.1016/S0960-9822(02)01073-4. PMID 12194817.
132. Matt Sullivan; Frank Uhlmann (2003). "A non-proteolytic function of separase links anaphase onset to mitotic exit". Nat Cell Biol. 5: 249–254. doi:10.1038/ncb940.
133. Rosella Visintin; Susanne Prinz; Angelika Amon (1997). "CDC20 and CDH1: A Family of Substrate-Specific Activators of APC-Dependent Proteolysis". Science. 278 (5337): 460–463. doi:10.1126/science.278.5337.460. PMID 9334304.
134. Steven I. Reed (2003). "Ratchets and clocks: the cell cycle, ubiquitylation and protein turnover". Nature Reviews Molecular Cell Biology. 4: 855–864. doi:10.1038/nrm1246.
135. P. K. Vinod; Paula Freire; Ahmed Rattani; Andrea Ciliberto; Frank Uhlmann & Bela Novak (2011). Computational modeling of mitotic exit in budding yeast: the role of separase and Cdc14 endocycles. J.R.Soc.Interface. 8: 1128–1141. doi:10.1098/rsif.2010.0649.
136. Tamara A. Potapova; John R. Daum; Bradley D. Pittman; Joanna R. Hudson; Tara N. Jones; David L. Satinover; P. Todd Stukenberg & Gary J. Gorbsky (2006). "The reversibility of mitotic exit in vertebrate cells". Nature Letters. 440: 954–958. doi:10.1038/nature04652.
137. Strogatz, Steven H, ed. (1994). Chapter 2 and 3. Nonlinear dynamics and chaos : with applications to physics, biology, chemistry, and engineering. Perseus Books.
138. John J. Tyson; Attila Csikasz-Nagy & Bela Novak (2002). "The dynamics of cell cycle regulation". BioEssays. 24 (12): 1095–1109. doi:10.1002/bies.10191.
139. Dan Siegal-Gaskins; Maria Katherine Mejia-Guerra; Gregory D. Smith; Erich Grotewold (2011). "Emergence of Switch-Like Behavior in a Large Family of Simple Biochemical Networks". PLOS Computational Biology. 7 (5): 1–12. doi:10.1371/journal.pcbi.1002039.
140. Crawford, John (1991). "Introduction to Bifurcation Theory". Reviews of Modern Physics. 63: 991–1037. doi:10.1103/revmodphys.63.991.
141. Visintin R, Hwang ES, Amon A (1999). "Cfi1 prevents premature exit from mitosis by anchoring Cdc14 phosphatase in the nucleolus". Nature. 398: 818–823. doi:10.1038/19775.

142. Lindqvist; W. van Zon; Rosenthal C. Karlsson; RM. Wolthuis (2007). Cyclin B1–Cdk1 Activation Continues after Centrosome Separation to Control Mitotic Progression. PLOS Biology. 5 (5): 1127–1137. doi:10.1371/journal.pbio.0050123.

143. Joanna Bloom; Frederick R. Cross (2007). Multiple levels of cyclin specificity in cell-cycle control. Nature Reviews Molecular Cell Biology. 8 (2): 149–160. doi:10.1038/nrm2105. PMID 17245415.

144. Charvin G, Oikonomou C, Siggia ED, Cross FR (2010). "Origin of Irreversibility of Cell Cycle Start in Budding Yeast". PLOS Biology. 8 (1): 1–13.

CHAPTER 3

The Process of Mitosis

Chapter Contents

Mitosis is the process in which a eukaryotic cell nucleus splits in two, followed by division of the parent cell into two daughter cells. The word "mitosis" means "threads," and it refers to the threadlike appearance of chromosomes as the cell prepares to divide. Early microscopists were the first to observe these structures, and they also noted the appearance of a specialized network of microtubules during mitosis. These tubules, collectively known as the spindle, extend from structures called centrosomes — with one centrosome located at each of the opposite ends, or poles, of a cell. As mitosis progresses, the microtubules attach to the chromosomes, which have already duplicated their DNA and aligned across the center of the cell. The spindle tubules then shorten and move toward the poles of the cell. As they move, they pull the one copy of each chromosome with them to opposite poles of the cell. This process ensures that each daughter cell will contain one exact copy of the parent cell DNA.

INTRODUCTION

In cell biology, **mitosis** is a part of the cell cycle when replicated chromosomes are separated into two new nuclei. In general, mitosis (division of the nucleus) is preceded by the S stage of interphase (during which the DNA is replicated) and is often accompanied or followed by cytokinesis, which divides the cytoplasm, organelles and cell membrane into two new cells containing roughly equal shares of these cellular components. Mitosis and cytokinesis together define the **mitotic (M) phase** of an animal cell cycle—the division of the mother cell into two daughter cells genetically identical to each other.

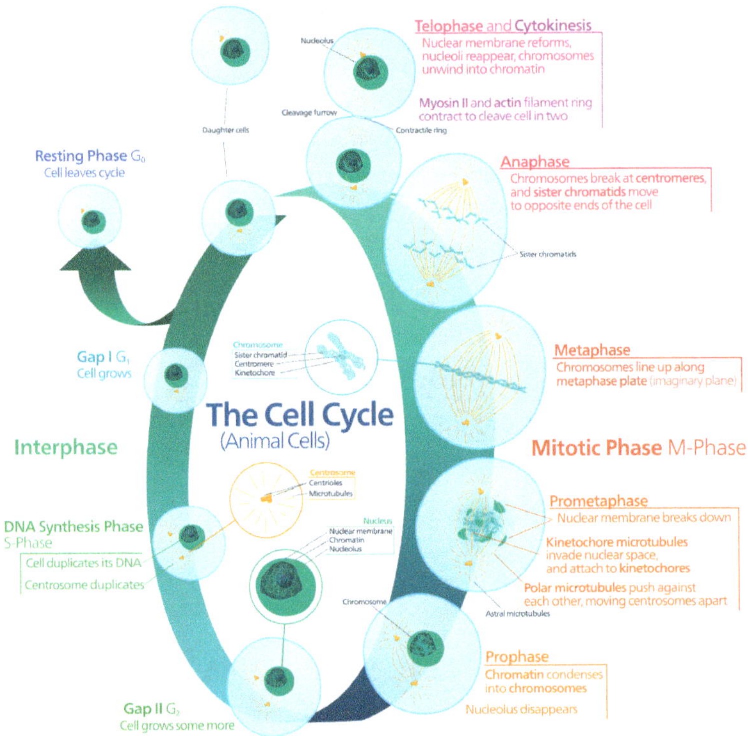

Mitosis in an animal cell (phases ordered counter-clockwise).

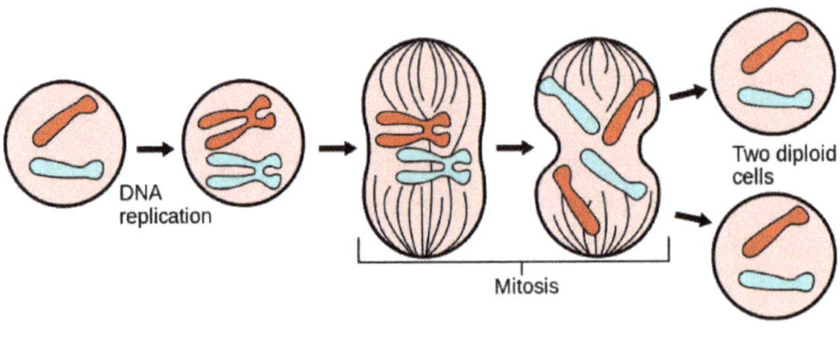

Mitosis divides the chromosomes in a cell nucleus.

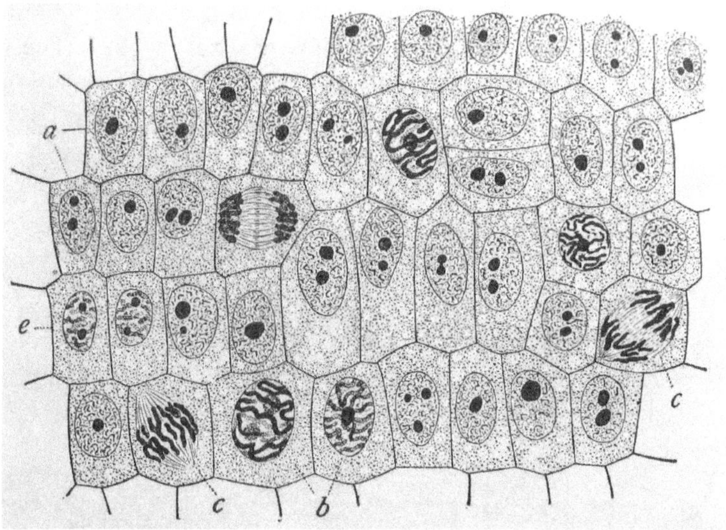

Onion (Allium) cells in different phases of the cell cycle enlarged 800 diameters.

a. non-dividing cells
b. nuclei preparing for division (spireme-stage)
c. dividing cells showing mitotic figures
e. pair of daughter-cells shortly after division.

The process of mitosis is divided into stages corresponding to the completion of one set of activities and the start of the next. These stages are prophase, prometaphase, metaphase, anaphase, and telophase. During mitosis, the chromosomes, which have already duplicated, condense and attach to spindle fibers that pull one copy of each chromosome to opposite sides of the cell. The result is two genetically identical daughter nuclei. The rest of the cell may then continue to divide by cytokinesis to produce two daughter cells. Producing three or more daughter cells instead of normal two is a mitotic error called tripolar mitosis or multipolar mitosis (direct cell triplication / multiplication). Other errors during mitosis can induce apoptosis (programmed cell death) or cause mutations. Certain types of cancer can arise from such mutations.

Mitosis occurs only in eukaryotic cells and the process varies in different organisms. For example, animals undergo an "open" mitosis, where the nuclear envelope breaks down before the chromosomes separate, while fungi undergo a "closed" mitosis, where chromosomes divide within an intact cell nucleus. Furthermore, most animal cells undergo a shape change, known as mitotic cell rounding, to adopt a near spherical morphology at the start of mitosis. Prokaryotic cells, which lack a nucleus, divide by a different process called binary fission.

DISCOVERY

German zoologist Otto Bütschli might have claimed the discovery of the process presently known as "mitosis", a term coined by Walther Flemming in 1882.

Mitosis was discovered in frog, rabbit, and cat cornea cells in 1873 and described for the first time by the Polish histologist Wacław Mayzel in 1875. The term is derived from the Greek word μίτος *mitos* "warp thread".

OVERVIEW

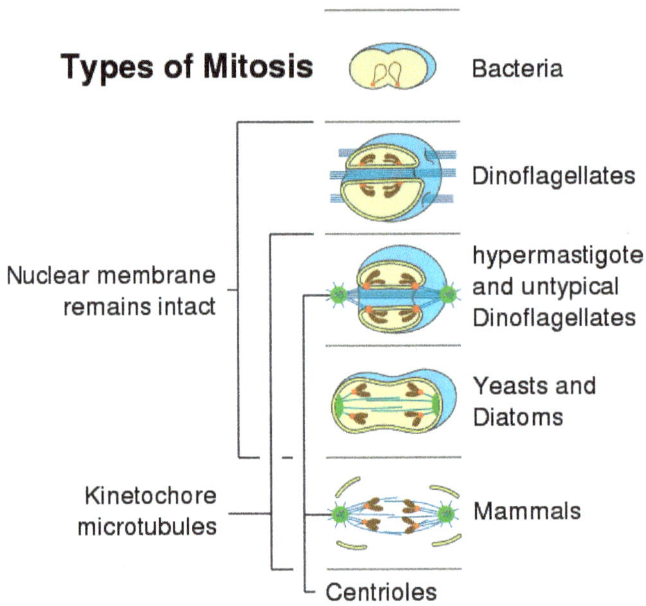

TYPES OF MITOSIS

The primary result of mitosis and cytokinesis is the transfer of a parent cell's genome into two daughter cells. The genome is composed of a number of chromosomes—complexes of tightly coiled DNA that contain genetic information vital for proper cell function. Because each resultant daughter cell should be genetically identical to the parent cell, the parent cell must make a copy of each chromosome before mitosis. This occurs during the S phase of interphase. Chromosome duplication results in two identical *sister chromatids* bound together by cohesin proteins at the *centromere.*

When mitosis begins, the chromosomes condense and become visible. In some eukaryotes, for example animals, the nuclear envelope, which segregates the DNA from the cytoplasm, disintegrates into small vesicles. The nucleolus, which makes ribosomes in the cell, also disappears. Microtubules project from opposite ends of the cell, attach to the centromeres, and align the chromosomes centrally within the cell. The microtubules then contract to pull the sister chromatids of each chromosome apart. Sister chromatids at this point are called *daughter chromosomes*. As the cell elongates, corresponding daughter chromosomes are pulled toward opposite ends of the cell and condense maximally in late anaphase. A new nuclear envelope forms around the separated daughter chromosomes, which decondense to form interphase nuclei.

During mitotic progression, typically after the anaphase onset, the cell may undergo cytokinesis. In animal cells, a cell membrane pinches inward between the two developing nuclei to produce two new cells. In plant cells, a cell plate forms between the two nuclei. Cytokinesis does not always occur; coenocytic (a type of multinucleate condition) cells undergo mitosis without cytokinesis.

PHASES OF CELL CYCLE AND MITOSIS

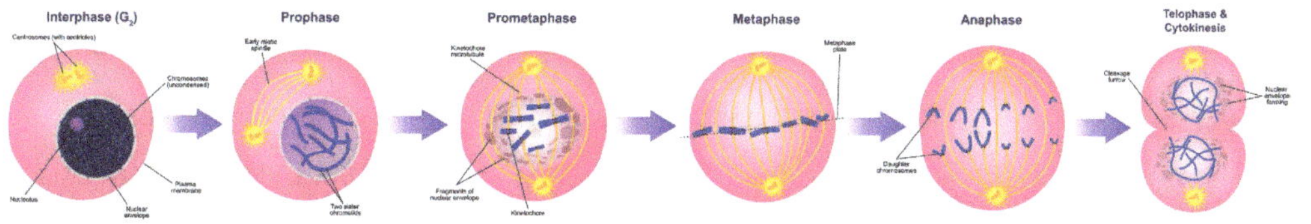

Interphase

The mitotic phase is a relatively short period of the cell cycle. It alternates with the much longer *interphase*, where the cell prepares itself for the process of cell division. Interphase is divided into three phases: G_1 (first gap), S (synthesis), and G_2 (second gap). During all three phases, the cell grows by producing proteins and cytoplasmic organelles. However, chromosomes are replicated only during the S phase. Thus, a cell grows (G_1), continues to grow as it duplicates its chromosomes (S), grows more and prepares for mitosis (G_2), and finally divides (M) before restarting the cycle. All these phases in the cell cycle are highly regulated by cyclins, cyclin-dependent kinases, and other cell cycle proteins. The phases follow one another in strict order and there are "checkpoints" that give the cell cues to proceed from one phase to another. Cells may also temporarily or permanently leave the cell cycle and enter G_0 phase to stop dividing. This can occur when cells become overcrowded (density-dependent inhibition) or when they differentiate to carry out specific functions for the organism, as is the case for human heart muscle cells and neurons. Some G_0 cells have the ability to re-enter the cell cycle.

Preprophase (plant cells)

In plant cells only, prophase is preceded by a pre-prophase stage. In highly vacuolated plant cells, the nucleus has to migrate into the center of the cell before mitosis can begin. This is achieved through the formation of a phragmosome, a transverse sheet of cytoplasm that bisects the cell along the future plane of cell division. In addition to phragmosome formation, preprophase is characterized by the formation of a ring of microtubules and actin filaments (called preprophase band) underneath the plasma membrane around the equatorial plane of the future mitotic spindle. This band marks the position where the cell will eventually divide. The cells of higher plants (such as the flowering plants) lack centrioles; instead, microtubules form a spindle on the surface of the nucleus and are then organized into a spindle by the chromosomes themselves, after the nuclear envelope breaks down. The preprophase band disappears during nuclear envelope breakdown and spindle formation in prometaphase.

Prophase

During prophase, which occurs after G_2 interphase, the cell prepares to divide by tightly condensing its chromosomes and initiating mitotic spindle formation, this process is called chromosome condensation. During interphase, the genetic material in the nucleus consists of loosely packed chromatin. At the onset of prophase, chromatin fibers condense into discrete chromosomes that are typically visible at high magnification through a light microscope. In this stage, chromosomes are long, thin and thread-like. Each chromosome has two chromatids. The two chromatids are joined at a place called centromere.

Gene transcription ceases during prophase and does not resume until late anaphase to early G1 phase. The nucleolus also disappears during early prophase.

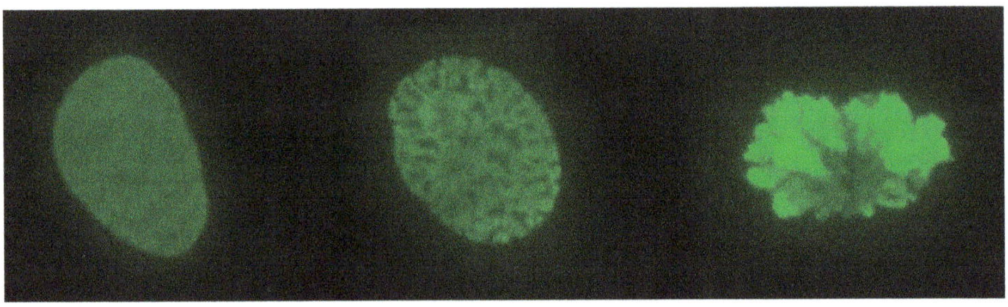

Condensing chromosomes. Interphase nucleus (left), condensing chromosomes (middle) and condensed chromosomes (right).

Close to the nucleus of animal cells are structures called centrosomes, consisting of a pair of centrioles surrounded by a loose collection of proteins. The centrosome is the coordinating center for the cell›s microtubules. A cell inherits a single centrosome at cell division, which is duplicated by the cell before a new round of mitosis begins, giving a pair of centrosomes. The two centrosomes polymerize tubulin to help form a microtubule spindle apparatus. Motor proteins then push the centrosomes along these microtubules to opposite sides of the cell. Although centrosomes help organize microtubule assembly, they are not essential for the formation of the spindle apparatus, since they are absent from plants, and are not absolutely required for animal cell mitosis.

Prometaphase

At the beginning of prometaphase in animal cells, phosphorylation of nuclear lamins causes the nuclear envelope to disintegrate into small membrane vesicles. As this happens, microtubules invade the nuclear space. This is called *open mitosis*, and it occurs in some multicellular organisms. Fungi and some protists, such as algae or trichomonads, undergo a variation called *closed mitosis* where the spindle forms inside the nucleus, or the microtubules penetrate the intact nuclear envelope.

In late prometaphase, *kinetochore microtubules* begin to search for and attach to chromosomal kinetochores. A *kinetochore* is a proteinaceous microtubule-binding structure that forms on the chromosomal centromere during late prophase. A number of *polar microtubules* find and interact with corresponding polar microtubules from the opposite centrosome to form the mitotic spindle. Although the kinetochore structure and function are not fully understood, it is known that it contains some form of molecular motor. When a microtubule connects with the kinetochore, the motor activates, using energy from ATP to "crawl" up the tube toward the originating centrosome. This motor activity, coupled with polymerisation and depolymerisation of microtubules, provides the pulling force necessary to later separate the chromosome's two chromatids.

Metaphase

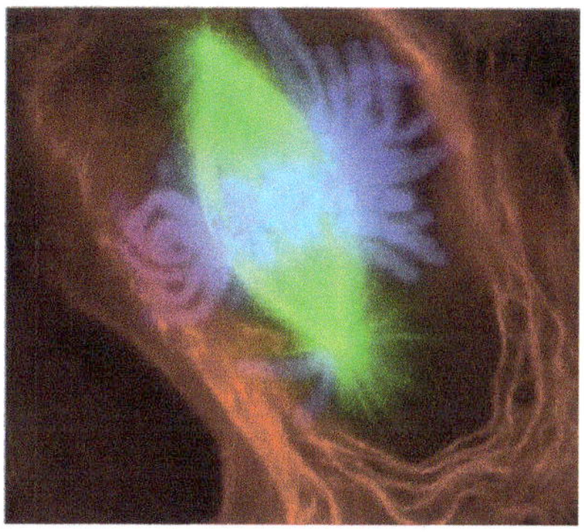

A cell in late metaphase. All chromosomes (blue) but one have arrived at the metaphase plate.

After the microtubules have located and attached to the kinetochores in prometaphase, the two centrosomes begin pulling the chromosomes towards opposite ends of the cell. The resulting tension causes the chromosomes to align along the *metaphase plate* or *equatorial plane*, an imaginary line that is centrally located between the two centrosomes (at approximately the midline of the cell). To ensure equitable distribution of chromosomes at the end of mitosis, the *metaphase checkpoint* guarantees that kinetochores are properly attached to the mitotic spindle and that the chromosomes are aligned along the metaphase plate. If the cell successfully passes through the metaphase checkpoint, it proceeds to anaphase.

Anaphase

During *anaphase A*, the cohesins that bind sister chromatids together are cleaved, forming two identical daughter chromosomes. Shortening of the kinetochore microtubules pulls the newly formed daughter chromosomes to opposite ends of the cell. During *anaphase B*, polar microtubules push against each other, causing the cell to elongate. In late anaphase, chromosomes also reach their overall maximal condensation level, to help chromosome segregation and the re-formation of the nucleus. In most animal cells, anaphase A precedes anaphase B, but some vertebrate egg cells demonstrate the opposite order of events.

Telophase

Telophase (from the Greek word *τελος* meaning end) is a reversal of prophase and prometaphase events. At telophase, the polar microtubules continue to lengthen, elongating the cell even more. If the nuclear envelope has broken down, a new nuclear envelope forms using the membrane vesicles of the parent cell›s old nuclear envelope. The new envelope forms around each set of separated daughter chromosomes (though the membrane does not enclose the centrosomes) and the nucleolus reappears. Both sets of chromosomes, now surrounded by new nuclear membrane, begin to relax or decondense. Mitosis is complete. Each daughter nucleus has an identical set of chromosomes. Cell division may or may not occur at this time depending on the organism.

CYTOKINESIS

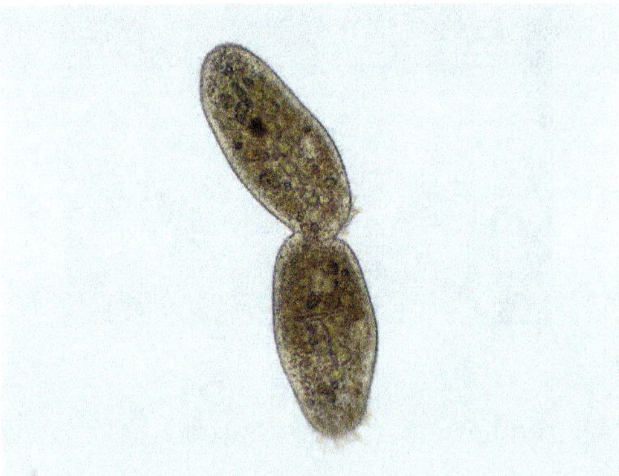

Cilliate undergoing cytokinesis, with the cleavage furrow being clearly visible

Cytokinesis is that part of the cell division process during which the cytoplasm of a single eukaryotic cell divides into two daughter cells. Cytoplasmic division begins during or after the late stages of nuclear division in mitosis and meiosis. During cytokinesis the spindle apparatus partitions and transports duplicated chromatids into the cytoplasm of the separating daughter cells. It thereby ensures that chromosome number and complement are maintained from one generation to the next and that except in special cases the daughter cells will be functional copies of the parent cell. After the completion of the telophase and cytokinesis each daughter cell enters the interphase of the cell cycle.

Particular functions demand various deviations from the process of symmetrical cytokinesis; for example in oogenesis in animals the ovum takes almost all the cytoplasm and organelles. This leaves very little for the resulting polar bodies, which in most species die without function, though they do take on various special functions in other species. Another form of mitosis occurs in tissues such as liver and skeletal muscle; it omits cytokinesis, thereby yielding multinucleate cells.

Plant cytokinesis differs from animal cytokinesis, partly because of the rigidity of plant cell walls. Instead of plant cells forming a cleavage furrow such as develops between animal daughter cells, a dividing structure known as the cell plate forms in the cytoplasm and grows into a new, doubled cell wall between plant daughter cells.

Cytokinesis largely resembles the prokaryotic process of binary fission, but because of differences between prokaryotic and eukaryotic cell structures and functions, the mechanisms differ. For instance a bacterial cell has only a single chromosome in the form of a closed loop, in contrast to the linear, usually multiple, chromosomes of eukaryotes; accordingly bacteria construct no mitotic spindle in cell division. Also, duplication of prokaryotic DNA takes place during the actual separation of chromosomes; in mitosis duplication takes place during the interphase before mitosis begins, though the daughter chromatids do not separate completely before the anaphase.

Animal Cell Cytokinesis

Animal cell cytokinesis begins shortly after the onset of sister chromatid separation in the anaphase of mitosis. The process can be divided to the following distinct steps: anaphase spindle reorganization, division plane specification, actin-myosin ring assembly and contraction, and abscission. Faithful partitioning of the genome to emerging daughter cells is ensured through the tight temporal coordination of the above individual events by molecular signaling pathways.

Anaphase Spindle Reorganization

Animal cell cytokinesis starts with the stabilization of microtubules and reorganization of the mitotic spindle to form the central spindle. central spindle (or *spindle midzone*) forms when non-kinetochore microtubule fibres are bundled between the spindle poles. A number of different species including *H. sapiens*, *D. melanogaster* and *C. elegans* require the central spindle in order to efficiently undergo cytokinesis, although the specific phenotype described when it is absent varies from one species to the next (for example, certain Drosophila cell types are incapable of forming a cleavage furrow without the central spindle, whereas in both *C. elegans* embryos and human tissue culture cells a cleavage furrow is observed to form and ingress, but then regress before cytokinesis is complete). The process of mitotic spindle reorganization and central spindle formation is caused by the decline of CDK1 activity during anaphase. The decline of CDK1 activity at the metaphase-anaphase transition leads to dephosphorylating of inhibitory sites on multiple central spindle components. First of all, the removal of a CDK1 phosphorylation from a subunit of the CPC (the chromosomal passenger complex) allows its translocalization to the central spindle from the centromeres, where it is located during metaphase. Besides being a structural component of the central spindle itself, CPC also plays a role in the phosphoregulation of other central spindle components, including PRC1 (microtubule bundling protein required for cytokinesis 1) and MKLP1 (a kinesin motor protein). Originally inhibited by CDK1-mediated phosphorylation, PRC1 is now able to form a homodimer that selectively binds to the interface between antiparallel microtubules, facilitating spatial organization of the microtubules of the central spindle. MKLP1, together with the Rho-family GTPase activating protein CYK-4 (also termed MgcRacGAP), forms the centralspindlin complex. Centralspindlin binds to the central spindle as higher-order clusters. The centralspindlin cluster formation is promoted by phosphorylation of MLKP1 by Aurora B, a component of CPC. In short, the self-assembly of central spindle is initiated through the phosphoregulation of multiple central spindle components by the decline of CDK1 activity, either directly or indirectly, at the metaphase–anaphase transition. The central spindle may have multiple functions in cytokinesis including the control of cleavage furrow positioning, the delivery of membrane vesicles to the cleavage furrow, and the formation of the midbody structure that is required for the final steps of division.

Division Plane Specification

The second step of animal cell cytokinesis involves division plane specification and cytokinetic furrow formation. Precise positioning of the division plane between the two masses of segregated chromosomes is essential to prevent chromosome loss. Meanwhile, the mechanism by which the spindle determines the division plane in animal cells is perhaps the most enduring mystery in cytokinesis and a matter of intense debate. There exist three hypotheses of furrow induction. The first is the astral stimulation hypothesis, which postulates that astral microtubules from the spindle poles carry a furrow-inducing signal to the cell cortex, where signals from two poles are somehow focused into a ring at the spindle. A second possibility, called the central spindle hypothesis, is that the cleavage furrow is induced by a positive stimulus that originates in the central spindle equator. The central spindle may contribute to the specification of the division plane by promoting concentration and activation of the small GTPase RhoA at the equatorial cortex. A third hypothesis is the astral relaxation hypothesis. It proposes that postulates that active actin–myosin bundles are distributed throughout the cell cortex, and inhibition of their contraction near the spindle poles results in a gradient of contractile activity that is highest at the midpoint between poles. In other words, astral microtubules generate a negative signal that increases cortical relaxation close to the poles. Genetic and laser-micromanipulation studies in C. elegans embryos haven shown that the spindle sends two redundant signals to the cell cortex, one originating from the central spindle, and a second signal deriving from the spindle aster, suggesting the involvement of multiple mechanisms combined in the positioning of the cleavage furrow. The predominance of one particular signal varies between cell types and organisms. And the multitude and partial redundancy of signals may be required to make the system robust and to increase spatial precision.

Actin-Myosin Ring Assembly and Contraction

At the cytokinesis furrow, it is the actin-myosin contractile ring that drives the cleavage process, during which cell membrane and wall grow inward, which eventually pinches the mother cell in two. The key components of this ring are the filamentous protein actin and the motor protein myosin II. The contractile ring assembles equatorially (in the middle of the cell) at the cell cortex (adjacent to the cell membrane). Rho protein family (RhoA protein in mammalian cells) is a key regulator of contractile ring formation and contraction in animal cells. The RhoA pathway promotes assembly of the actin-myosin ring by two main effectors. First, RhoA stimulates nucleation of unbranched actin filaments by activation of Diaphanous-related formins. This local generation of new actin filaments is important for the contractile ring formation. This actin filament formation process also requires a protein called profilin, which binds to actin monomers and helps load them onto the filament end. Second, RhoA promotes myosin II activation by the kinase ROCK, which activates myosin II directly by phosphorylation of the myosin light chain and also inhibits myosin phosphatase by phosphorylation of the phosphatase-targeting subunit MYPT. Besides actin and myosin II, the contractile ring contains the scaffolding protein anillin. Anillin binds to actin, myosin, RhoA and CYK-4, and thereby links the equatorial cortex with the signals from the central spindle. It also contributes to the linkage of the actin-myosin ring to the plasma membrane. Another protein, septin, has also been speculated to serve as a structural scaffold on which the cytokinesis apparatus is organized. Following its assembly, contraction of the actin-myosin ring leads to ingression of the attached plasma membrane, which partitions the cytoplasm into two domains of emerging sister cells. The force for the contractile processes is generated by movements along actin by the motor protein myosin II. Myosin II uses the free energy released when ATP is hydrolysed to move along these actin filaments, constricting the cell membrane to form a cleavage furrow. Continued hydrolysis causes this cleavage furrow to ingress (move inwards), a striking process that is clearly visible through a light microscope.

Abscission

The cytokinetic furrow ingresses until a so-called midbody structure (composed of electron-dense, proteinaceous material) is formed, where the actin-myosin ring has reached a diameter of about 1–2 μm. Most animal cell types remain connected by an intercellular bridge for up to several hours until they are split by an actin-independent process termed abscission, the last step of cytokinesis. The process of abscission physically cleaves the aforementioned midbody into two. Abscission proceeds by removal of cytoskeletal structures from the intercellular bridge, constriction of the cell cortex, and plasma membrane fission. The intercellular bridge is filled with dense bundles of antiparallel microtubules that derive from the central spindle. These microtubules overlap at the midbody, which is generally thought as a targeting platform for the abscission machinery. Microtubule severing protein spastin is largely responsible for the disassembly of microtubule bundles inside the intercellular bridge. Complete cortical constriction also requires removal of the underlying cytoskeletal structures. Actin filament disassembly during late cytokinesis depends on the PKCε–14-3-3 complex, which inactivates RhoA after furrow ingression. Actin disassembly is further controlled by the GTPase Rab35 and its effector, the phosphatidylinositol-4,5-bisphosphate 5-phosphatase OCRL. Understanding the mechanism by which the plasma membrane ultimately splits requires further investigation.

Timing Cytokinesis

Cytokinesis must be temporally controlled to ensure that it occurs only after sister chromatids separate during the anaphase portion of normal proliferative cell divisions. To achieve this, many components of the cytokinesis machinery are highly regulated to ensure that they are able to perform a particular function at only a particular stage of the cell cycle. Cytokinesis happens only after APC binds with CDC20. This allows for the separation of chromosomes and myosin to work simultaneously.

After cytokinesis, non-kinetochore microtubules reorganize and disappear into a new cytoskeleton as the cell cycle returns to interphase (see also cell cycle).

Plant Cell Cytokinesis

Due to the presence of a cell wall, cytokinesis in plant cells is significantly different from that in animal cells, Rather than forming a contractile ring, plant cells construct a cell plate in the middle of the cell. The stages of cell plate formation include (1) creation of the phragmoplast, an array of microtubules that guides and supports the formation of the cell plate; (2) trafficking of vesicles to the division plane and their fusion to generate a tubular-vesicular network; (3) continued fusion of membrane tubules and their transformation into membrane sheets upon the deposition of callose, followed by deposition of cellulose and other cell wall components; (4) recycling of excess membrane and other material from the cell plate; and (5) fusion with the parental cell wall

The phragmoplast is assembled from the remnants of the mitotic spindle, and serves as a track for the trafficking of vesicles to the phragmoplast midzone. These vesicles contain lipids, proteins and carbohydrates needed for the formation of a new cell boundary. Electron tomographic studies have identified the Golgi apparatus as the source of these vesicles, but other studies have suggested that they contain endocytosed material as well.

These tubules then widen and fuse laterally with each other, eventually forming a planar, fenestrated sheet . As the cell plate matures, large amounts of membrane material are removed via clathrin-mediated endocytosis Eventually, the edges of the cell plate fuse with the parental plasma membrane, often in an asymmetrical fashion, thus completing cytokinesis. The remaining fenestrae contain strands of endoplasmic reticulum passing through them, and are thought to be the precursors of plasmodesmata .

The construction of the new cell wall begins within the lumen of the narrow tubules of the young cell plate. The order in which different cell wall components are deposited has been determined largely by immuno-electron microscopy. The first components to arrive are pectins, hemicelluloses, and arabinogalactan proteins carried by the secretory vesicles that fuse to form the cell plate. The next component to be added is callose, which is polymerized directly at the cell plate by callose synthases. As the cell plate continues to mature and fuses with the parental plasma membrane, the callose is slowly replaced with cellulose, the primary component of a mature cell wall . The middle lamella (a glue-like layer containing pectin) develops from the cell plate, serving to bind the cell walls of adjoining cells together.

Forces in Cytokinesis

Animal cells

Cytokinetic furrow ingression is powered by Type II Myosin ATPase. Since Myosins are recruited to the medial region, the contractile forces acting on the cortex resemble a 'purse string' constriction pulling inwards. This leads to the inward constriction. The plasma membrane by virtue of its close association with the cortex via crosslinker proteins

Etymology and Pronunciation

The word *cytokinesis* uses combining forms of *cyto-* + *kine-* + *-sis*, New Latin from Classical Latin and Ancient Greek, reflecting cell and *kinesis* ("motion, movement").

SIGNIFICANCE

Mitosis is important for the maintenance of the chromosomal set; each cell formed receives chromosomes that are alike in composition and equal in number to the chromosomes of the parent cell.

Mitosis occurs in the following circumstances:

Development and growth

The number of cells within an organism increases by mitosis. This is the basis of the development of a multicellular body from a single cell, i.e., zygote and also the basis of the growth of a multicellular body.

Cell replacement

In some parts of body, e.g. skin and digestive tract, cells are constantly sloughed off and replaced by new ones. New cells are formed by mitosis and so are exact copies of the cells being replaced. In like manner, red blood cells have short lifespan (only about 4 months) and new RBCs are formed by mitosis.

Regeneration

Some organisms can regenerate body parts. The production of new cells in such instances is achieved by mitosis. For example, starfish regenerate lost arms through mitosis.

Asexual reproduction

Some organisms produce genetically similar offspring through asexual reproduction. For example, the hydra reproduces asexually by budding. The cells at the surface of hydra undergo mitosis and form a mass called a bud. Mitosis continues in the cells of the bud and this grows into a new individual. The same division happens during asexual reproduction or vegetative propagation in plants.

CELL SHAPE CHANGES DURING MITOSIS

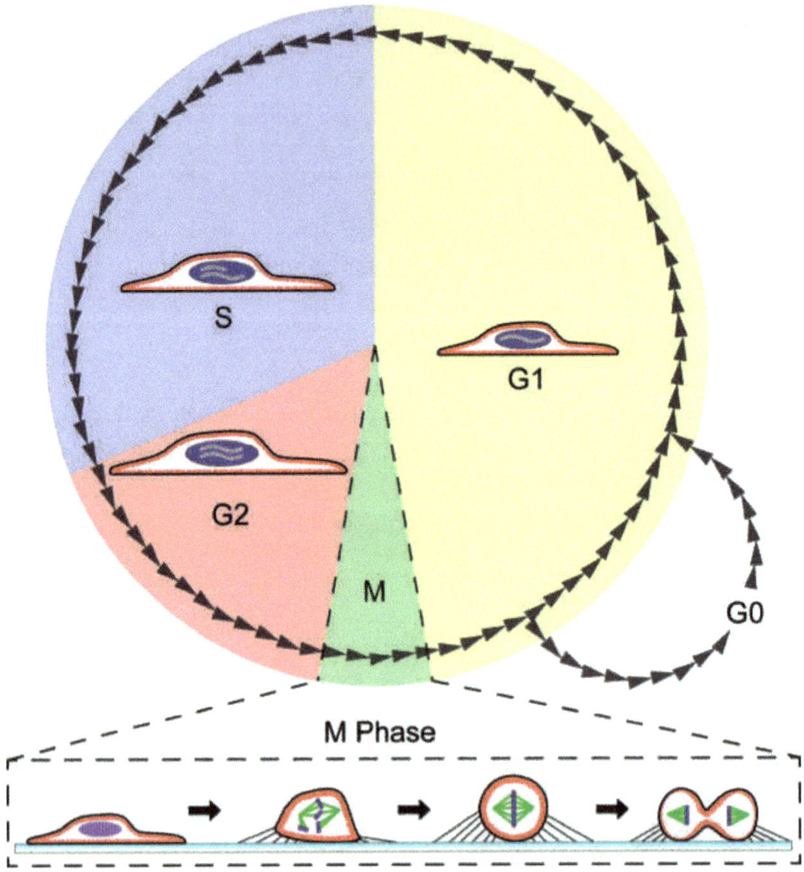

Cell shape changes through mitosis for a typical animal cell cultured on a flat surface. The cell undergoes mitotic cell rounding during spindle assembly and then divides via cytokinesis. The actomyosin cortex is depicted in red, DNA/chromosomes purple, microtubules green, and membrane and retraction fibers in black. Rounding also occurs in live tissue, as described in the text.

In animal tissue, most cells round up to a near-spherical shape during mitosis. In epithelia and epidermis, an efficient rounding process is correlated with proper mitotic spindle alignment and subsequent correct positioning of daughter cells. Moreover, researchers have found that if rounding is heavily suppressed it may result in spindle defects, primarily pole splitting and failure to efficiently capture chromosomes. Therefore, mitotic cell rounding is thought to play a protective role in ensuring accurate mitosis.

Rounding forces are driven by reorganization of F-actin and myosin (actomyosin) into a contractile homogeneous cell cortex that 1) rigidifies the cell periphery and 2) facilitates generation of intracellular hydrostatic pressure (up to 10 fold higher than interphase). The generation of intracellular pressure is particularly critical under confinement, such as would be important in a tissue scenario, where outward forces must be produced to round up against surrounding cells and/or the extracellular matrix. Generation of pressure is dependent on formin-mediated F-actin nucleation and Rho kinase (ROCK)-mediated myosin II contraction, both of which are governed upstream by signaling pathways RhoA and ECT2 through the activity of Cdk1. Due to its importance in mitosis, the molecular components and dynamics of the mitotic actomyosin cortex is an area of active research.

ERRORS AND VARIATIONS OF MITOSIS

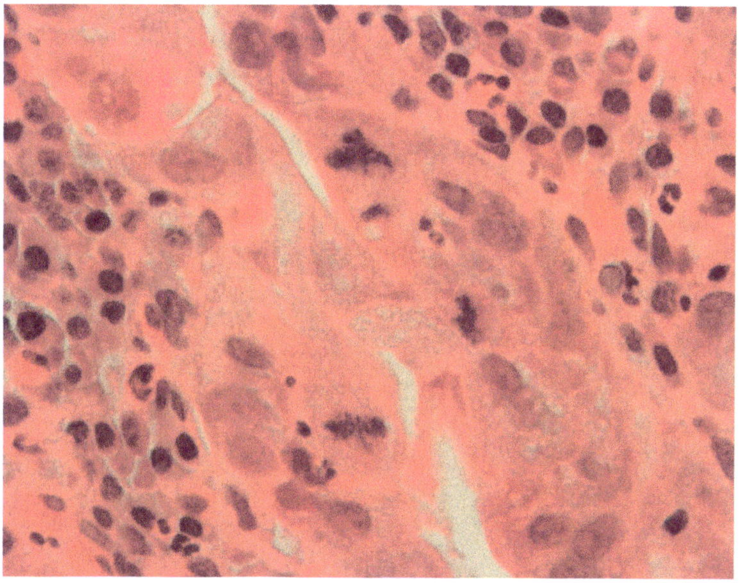

An abnormal (tripolar) mitosis (12 o'clock position) in a precancerous lesion of the stomach (H&E stain).

Errors can occur during mitosis, especially during early embryonic development in humans. Mitotic errors can create aneuploid cells that have too few or too many of one or more chromosomes, a condition associated with cancer. Early human embryos, cancer cells, infected or intoxicated cells can also suffer from pathological division into three or more daughter cells (tripolar or multipolar mitosis), resulting in severe errors in their chromosomal complements.

In *nondisjunction*, sister chromatids fail to separate during anaphase. One daughter cell receives both sister chromatids from the nondisjoining chromosome and the other cell receives none. As a result, the former cell gets three copies of the chromosome, a condition known as *trisomy*, and the latter will have only one copy, a condition known as *monosomy*. On occasion, when cells experience nondisjunction, they fail to complete cytokinesis and retain both nuclei in one cell, resulting in binucleated cells.

Anaphase lag occurs when the movement of one chromatid is impeded during anaphase. This may be caused by a failure of the mitotic spindle to properly attach to the chromosome. The lagging chromatid is excluded from both nuclei and is lost. Therefore, one of the daughter cells will be monosomic for that chromosome.

Endoreduplication (or endoreplication) occurs when chromosomes duplicate but the cell does not subsequently divide. This results in polyploid cells or, if the chromosomes duplicates repeatedly, polytene chromosomes. Endoreduplication is found in many species and appears to be a normal part of development. Endomitosis is a variant of endoreduplication in which cells replicate their chromosomes during S phase and enter, but prematurely terminate, mitosis. Instead of being divided into two new daughter nuclei, the replicated chromosomes are retained within the original nucleus. The cells then re-enter G_1 and S phase and replicate their chromosomes again. This may occur multiple times, increasing the chromosome number with each round of replication and endomitosis. Platelet-producing megakaryocytes go through endomitosis during cell differentiation.

Timeline in pictures

Mitotic cells can be visualized microscopically by staining them with fluorescent antibodies and dyes.

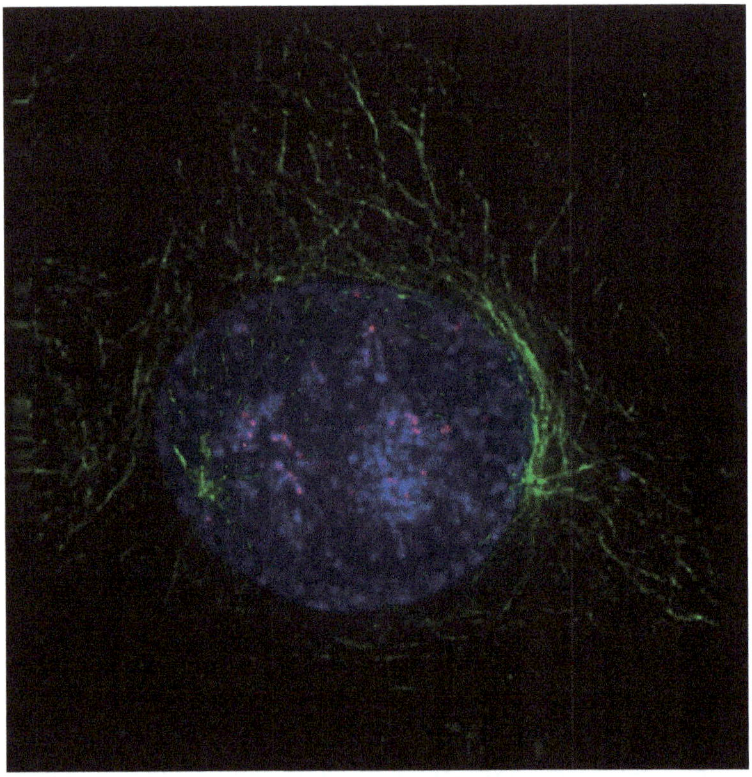

Early prophase: Nonkinetochore microtubules, shown as green strands, have established a matrix around the disassembling nucleus, with the condensing chromosomes in blue. The red nodules are the centromeres.

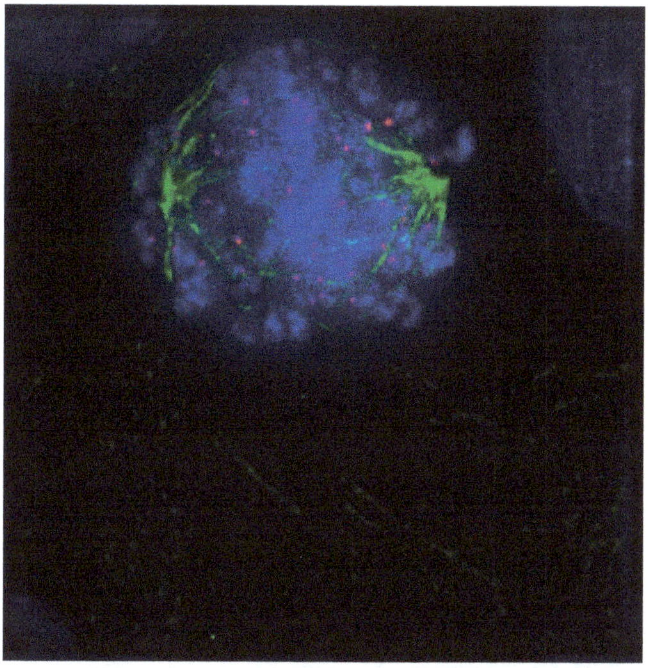

Early prometaphase: The nuclear envelope has just disassembled, allowing the microtubules to quickly interact with the kinetochores, which assemble on the centromeres of the condensing chromosomes.

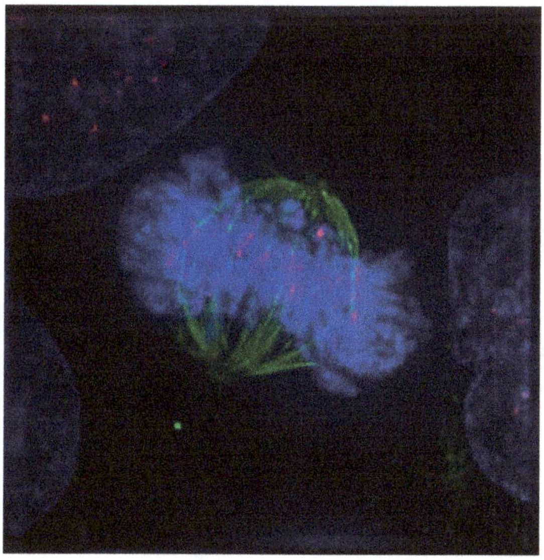

Metaphase: The centrosomes have moved to the poles of the cell and have established the mitotic spindle. The chromosomes have congressed at the metaphase plate.

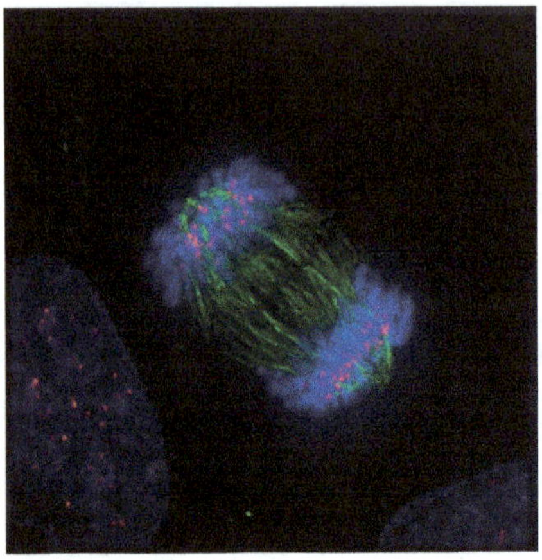

Anaphase: Kinetochore microtubules pull the two sets of chromosomes apart, and lengthening nonkinetochore microtubules push the halves of the dividing cell further apart, while chromosomes are condensed maximally.

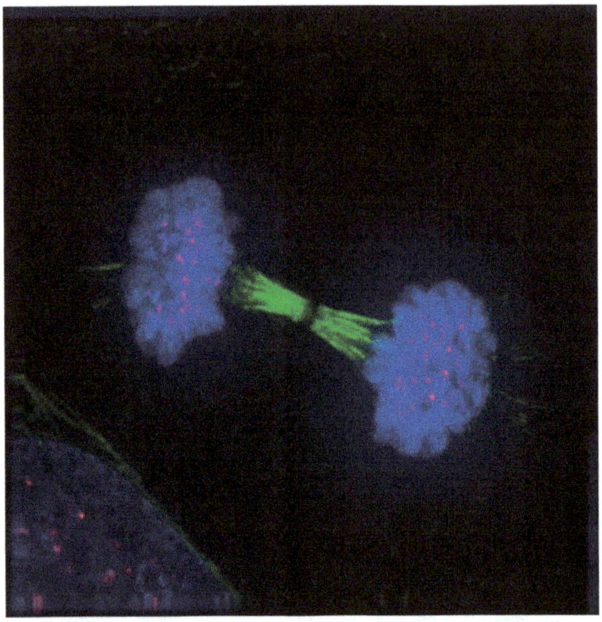

Telophase: Reversal of prophase and prometaphase events and thus completing the cell cycle.

CYCLIN

Cyclins are a family of proteins that control the progression of cells through the cell cycle by activating cyclin-dependent kinase (Cdk) enzymes.

Etymology

Cyclins were originally discovered by R. Timothy Hunt in 1982 while studying the cell cycle of sea urchins.

In an interview for The Life Scientific (aired on 13/12/2011) hosted by Jim Al-Khalili, R. Timothy Hunt explained that the name "cyclin" was originally named after his hobby cycling. It was only after the naming did its importance in the cell cycle become apparent. As it was appropriate the name stuck. R. Timothy Hunt: By the way, the name cyclin, which I coined, was really a joke, it›s because I liked cycling so much at the time, but they did come and go in the cell...

Function

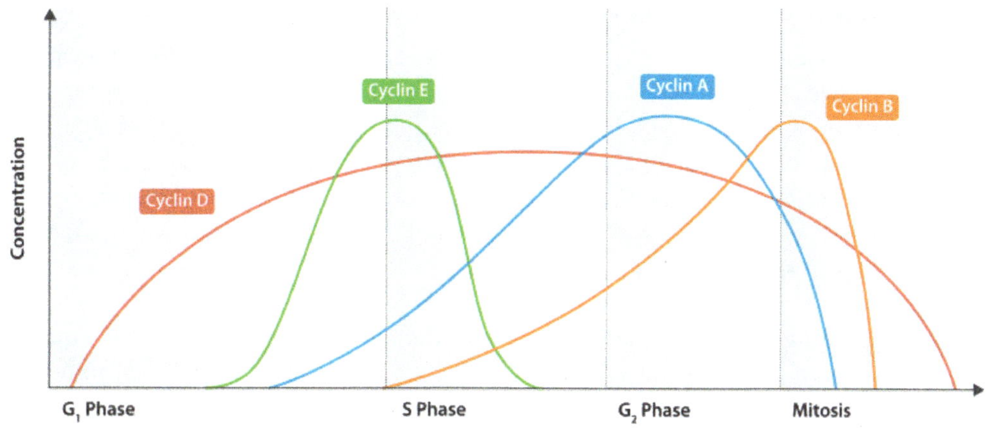

Expression of human cyclins through the cell cycle.

Cyclins were originally named because their concentration varies in a cyclical fashion during the cell cycle. (Note that the cyclins are now classified according to their conserved cyclin box structure, and not all these cyclins alter in level through the cell cycle.) The oscillations of the cyclins, namely fluctuations in cyclin gene expression and destruction by the ubiquitin mediated proteasome pathway, induce oscillations in Cdk activity to drive the cell cycle. A cyclin forms a complex with Cdk, which begins to activate the Cdk, but the complete activation requires phosphorylation, as well. Complex formation results in activation of the Cdk active site. Cyclins themselves have no enzymatic activity but have binding sites for some substrates and target the Cdks to specific subcellular locations.

Cyclins, when bound with the dependent kinases, such as the p34/cdc2/cdk1 protein, form the maturation-promoting factor. MPFs activate other proteins through phosphorylation. These phosphorylated proteins, in turn, are responsible for specific events during cycle division such as microtubule formation and chromatin remodeling. Cyclins can be divided into four classes based on their behavior in the cell cycle of vertebrate somatic cells and yeast cells: G1/S cyclins, S cyclins, G2 cyclins, M cyclins. This division is useful when talking about most cell cycles, but it is not universal as some cyclins have different functions or timing in different cell types.

G1/S Cyclins rise in late G1 and fall in early S phase. The Cdk- G1/S cyclin complex begins to induce the initial processes of DNA replication, primarily by arresting systems that prevent S phase Cdk activity in G1. The cyclins also promote other activities to progress the cell cycle, such as centrosome duplication in vertebrates or spindle pole body in yeast. The rise in presence of G1/S cyclins is paralleled by a rise in S cyclins.

G1 cyclins do not behave like the other cyclins, in that the concentrations increase gradually (with no oscillation), throughout the cell cycle based on cell growth and the external growth-regulatory signals. The presence of G cyclins coordinate cell growth with the entry to a new cell cycle.

S cyclins bind to Cdk and the complex directly induces DNA replication. The levels of S cyclins remain high, not only throughout S phase, but through G2 and early mitosis as well to promote early events in mitosis.

M cyclin concentrations rise as the cell begins to enter mitosis and the concentrations peak at metaphase. Cell changes in the cell cycle like the assembly of mitotic spindles and alignment of sister-chromatids along the spindles are induced by M cyclin- Cdk complexes. The destruction of M cyclins during metaphase and anaphase, after the Spindle Assembly Checkpoint is satisfied, causes the exit of mitosis and cytokinesis.

Domain Structure

Cyclins are generally very different from each other in primary structure, or amino acid sequence. However, all members of the cyclin family are similar in 100 amino acids that make up the cyclin box. Cyclins contain two domains of a similar all-α fold, the first located at the N-terminus and the second at the C-terminus. All cyclins are believed to contain a similar tertiary structure of two compact domains of 5 α helices. The first of which is the conserved cyclin box, outside of which cyclins are divergent. For example, the amino-terminal regions of S and M cyclins contain short destruction-box motifs that target these proteins for proteolysis in mitosis.

Cyclin, N-terminal domain

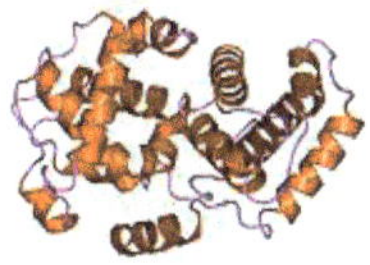

Structure of bovine cyclin A.

Identifiers	
Symbol	Cyclin_N
Pfam	PF00134
Pfam clan	CL0065
InterPro	IPR006671
PROSITE	PDOC00264
SCOP	1vin
SUPERFAMILY	1vin

Cyclin, C-terminal domain

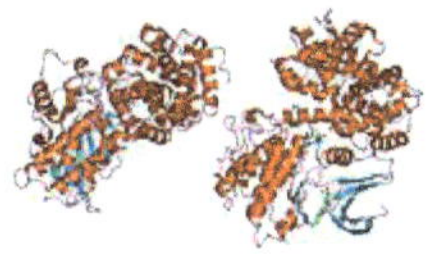

Structure of CDK2-cyclin A/ indirubin-5-sulphonate.

Identifiers	
Symbol	Cyclin_C
Pfam	PF02984
Pfam clan	CL0065
InterPro	IPR004367
PROSITE	PDOC00264
SCOP	1vin
SUPERFAMILY	1vin

K cyclin, C terminal

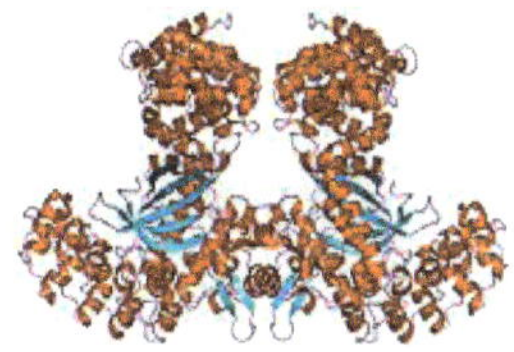

structure of a p18(ink4c)-cdk6-k-cyclin ternary complex

Identifiers	
Symbol	K-cyclin_vir_C
Pfam	PF09080
InterPro	IPR015164
SCOP	1g3n
SUPERFAMILY	1g3n

Types

There are several different cyclins that are active in different parts of the cell cycle and that cause the Cdk to phosphorylate different substrates. There are also several "orphan" cyclins for which no Cdk partner has been identified. For example, cyclin F is an orphan cyclin that is essential for G_2/M transition. A study in *C. elegans* revealed the specific roles of mitotic cyclins. Notably, recent studies have shown that cyclin A creates a cellular environment that promotes microtubule detachment from kinetochores in prometaphase to ensure efficient error correction and faithful chromosome segregation. Cells must separate their chromosomes precisely, an event that relies on the bi-oriented attachment of chromosomes to spindle microtubules through specialized structures called kinetochores. In the early phases of division, there are numerous errors in how kinetochores bind to spindle microtubules. The unstable attachments promote the correction of errors by causing a constant detachment, realignment and reattachment of microtubules from kinetochores in the cells as they try to find the correct attachment. Protein cyclin A governs this process by keeping the process going until the errors are eliminated. In normal cells, persistent cyclin A expression prevents the stabilization of microtubules bound to kinetochores even in cells with aligned chromosomes. As levels of cyclin A decline, microtubule attachments become stable, allowing the chromosomes to be divided correctly as cell division proceeds. In contrast, in cyclin A-deficient cells, microtubule attachments are prematurely stabilized. Consequently, these cells may fail to correct errors, leading to higher rates of chromosome mis-segregation.

Main Groups

There are two main groups of cyclins:

- G_1/S cyclins – essential for the control of the cell cycle at the G_1/S transition,
    - Cyclin A / CDK2 – active in S phase.
    - Cyclin D / CDK4, Cyclin D / CDK6, and Cyclin E / CDK2 – regulates transition from G_1 to S phase.
- G_2/M cyclins – essential for the control of the cell cycle at the G2/M transition (mitosis). G_2/M cyclins accumulate steadily during G_2 and are abruptly destroyed as cells exit from mitosis (at the end of the M-phase).
    - Cyclin B / CDK1 – regulates progression from G_2 to M phase.

Subtypes

Specific cyclin subtypes include:

Species	G1	G1/S	S	M
S. cerevisiae	Cln3 (Cdk1)	Cln 1,2 (Cdk1)	Clb 5,6 (Cdk1)	Clb 1,2,3,4 (Cdk 1)
S. pombe	Puc1? (Cdk1)	Puc1, Cig1? (Cdk1)	Cig2, Cig1? (Cdk1)	Cdc13 (Cdk1)
D. melanogaster	cyclin D (Cdk4)	cyclin E (Cdk2)	cyclin E, A (Cdk2,1)	cyclin A, B, B3 (Cdk1)
X. laevis	either not known or not present	cyclin E (Cdk2)	cyclin E, A (Cdk2,1)	cyclin A, B, B3 (Cdk1)
H. sapiens	cyclin D 1,2,3 (Cdk4, Cdk6)	cyclin E (Cdk2)	cyclin A (Cdk2, Cdk1)	cyclin B (Cdk1)

family	members
A	CCNA1, CCNA2
B	CCNB1, CCNB2, CCNB3
C	CCNC
D	CCND1, CCND2, CCND3
E	CCNE1, CCNE2
F	CCNF
G	CCNG1, CCNG2
H	CCNH
I	CCNI, CCNI2
J	CCNJ, CCNJL
K	CCNK
L	CCNL1, CCNL2
O	CCNO
T	CCNT1, CCNT2
Y	CCNY, CCNYL1, CCNYL2, CCNYL3

Other Proteins Containing this Domain

In addition, the following human proteins contain a cyclin domain:

CABLES2, CNTD1, CNTD2

History

Leland H. Hartwell, R. Timothy Hunt, and Paul M. Nurse won the 2001 Nobel Prize in Physiology or Medicine for their discovery of cyclin and cyclin-dependent kinase.

CYCLIN-DEPENDENT KINASE

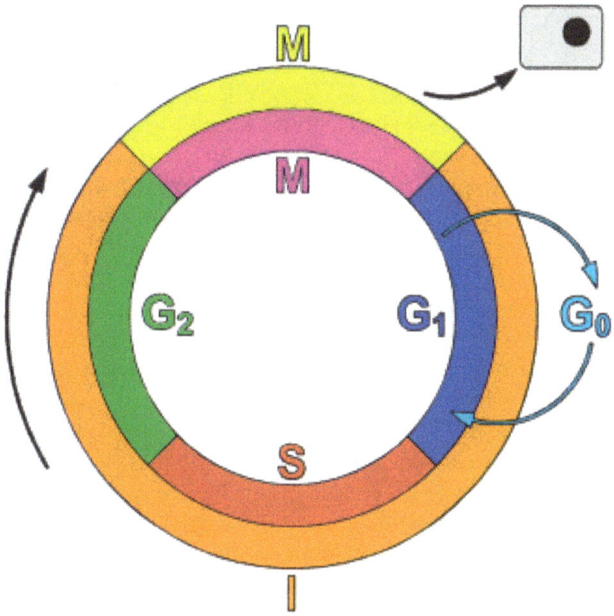

Schematic of the cell cycle. outer ring: I=Interphase, M=Mitosis; inner ring: M=Mitosis; G1=Gap phase 1; S=Synthesis; G2=Gap phase 2. The duration of mitosis in relation to the other phases has been exaggerated in this diagram.

Cyclin-dependent kinases (CDKs) are a family of protein kinases first discovered for their role in regulating the cell cycle. They are also involved in regulating transcription, mRNA processing, and the differentiation of nerve cells. They are present in all known eukaryotes, and their regulatory function in the cell cycle has been evolutionarily conserved. In fact, yeast cells can proliferate normally when their CDK gene has been replaced with the homologous human gene. CDKs are relatively small proteins, with molecular weights ranging from 34 to 40 kDa, and contain little more than the kinase domain. By definition, a CDK binds a regulatory protein called a cyclin. Without cyclin, CDK has little kinase activity; only the cyclin-CDK complex is an active kinase. CDKs phosphorylate their substrates on serines and threonines, so they are serine-threonine kinases. The consensus sequence for the phosphorylation site in the amino acid sequence of a CDK substrate is [S/T*]PX[K/R], where S/T* is the phosphorylated serine or threonine, P is proline, X is any amino acid, K is lysine, and R is arginine.

Types

Known CDKs, their cyclin partners, and their functions in the human and consequences of deletion in mice.

CDK	Cyclin partner	Function	Deletion Phenotype in Mice
Cdk1	Cyclin B	M phase	None. ~E2.5.
Cdk2	Cyclin E	G1/S transition	Reduced size, imparted neural progenitor cell proliferation. Viable, but both males & females sterile.
Cdk2	Cyclin A	S phase, G2 phase	
Cdk3	Cyclin C	G1 phase ?	No defects. Viable, fertile.
Cdk4	Cyclin D	G1 phase	Reduced size, insulin deficient diabetes. Viable, but both male & female infertile.
Cdk5	p35	Transcription	Severe neurological defects. Died immediately after birth.
Cdk6	Cyclin D	G1 phase	
Cdk7	Cyclin H	CDK-activating kinase, transcription	
Cdk8	Cyclin C	Transcription	Embryonic lethal
Cdk9	Cyclin T	Transcription	Embryonic lethal
Cdk11	Cyclin L	?	Mitotic defects. E3.5.
?	Cyclin F	?	Defects in extraembryonic tissues. E10.5.
?	Cyclin G	?	

CDKs and Cyclins in the Cell Cycle

Most of the known cyclin-CDK complexes regulate the progression through the cell cycle. Animal cells contain at least nine CDKs, four of which, CDK1, 2, 3, and 4, are directly involved in cell cycle regulation. In mammalian cells, CDK1, with its partners cyclin A2 and B1, alone can drive the cell cycle. Another one, CDK7, is involved indirectly as the CDK-activating kinase. Cyclin-CDK complexes phosphorylate substrates appropriate for the particular cell cycle phase. Cyclin-CDK complexes of earlier cell-cycle phase help activate cyclin-CDK complexes in later phase.

Cyclins and CDKs by Cell-Cycle Phase

Phase	Cyclin	CDK
G0	C	Cdk3
G1	D, E	Cdk4, Cdk2, Cdk6
S	A, E	Cdk2
G2	A	Cdk2, Cdk1
M	B	Cdk1

Cyclin-dependent kinases that control the cell cycle in model organisms.

Species	Name	Original name	Size (amino acids)	Function
Saccharomyces cerevisiae	Cdk1	Cdc28	298	All cell-cycle stages
Schizosaccharomyces pombe	Cdk1	Cdc2	297	All cell-cycle stages
Drosophila melanogaster	Cdk1	Cdc2	297	M
	Cdk2	Cdc2c	314	G1/S, S, possibly M
	Cdk4	Cdk4/6	317	G1, promotes growth
Xenopus laevis	Cdk1	Cdc2	301	M
	Cdk2		297	S, possibly M
Homo sapiens	Cdk1	Cdc2	297	M
	Cdk2		298	G1, S, possibly M
	Cdk4		301	G1
	Cdk6		326	G1

A list of CDKs with their regulator protein, cyclin or other.

- CDK1; cyclin A, cyclin B
- CDK2; cyclin A, cyclin E
- *CDK3*; cyclin C
- CDK4; cyclin D1, cyclin D2, cyclin D3
- CDK5; CDK5R1, CDK5R2. See also CDKL5.
- CDK6; cyclin D1, cyclin D2, cyclin D3
- CDK7; cyclin H
- *CDK8*; cyclin C
- CDK9; cyclin T1, cyclin T2a, cyclin T2b, cyclin K
- *CDK10*
- CDK11 (*CDC2L2*) ; cyclin L
- CDK12; cyclin L
- CDK13 (*CDC2L5*) ; cyclin L

Regulation of CDK Activity

CDK levels remain relatively constant throughout the cell cycle and most regulation is post-translational. Most knowledge of CDK structure and function is based on CDKs of *S. pombe* (Cdc2), *S. cerevisiae* (CDC28), and vertebrates (CDC2 and CDK2). The four major mechanisms of CDK regulation are cyclin binding, CAK phosphorylation, regulatory inhibitory phosphorylation, and binding of CDK inhibitory subunits (CKIs).

Cyclin Binding

The active site, or ATP-binding site, of all kinases is a cleft between a small amino-terminal lobe and a larger carboxy-terminal lobe. The structure of human Cdk2 revealed that CDKs have a modified ATP-binding site that can be regulated by cyclin binding. Phosphorylation by CDK-activating kinase (CAK) at Thr 161 on the T-loop increases the complex activity. Without cyclin, a flexible loop called the activation loop or T-loop blocks the cleft, and the position of several key amino acid residues is not optimal for ATP-binding. With cyclin, two alpha helices change position to permit ATP binding. One of them, the L12 helix that comes just before the T-loop in the primary sequence, becomes a beta strand and helps rearrange the T-loop, so it no longer blocks the active site. The other alpha helix called the PSTAIRE helix rearranges and helps change the position of the key amino acid residues in the active site.

There is considerable specificity in which cyclin binds with CDK. Furthermore, cyclin binding determines the specificity of the cyclin-CDK complex for particular substrates. Cyclins can directly bind the substrate or localize the CDK to a subcellular area where the substrate is found. Substrate specificity of S cyclins is imparted by the hydrophobic batch (centered on the MRAIL sequence), which has affinity for substrate proteins that contain a hydrophobic RXL (or Cy) motif. Cyclin B1 and B2 can localize Cdk1 to the nucleus and the Golgi, respectively, through a localization sequence outside the CDK-binding region.

Phosphorylation

Full kinase activity requires an activating phosphorylation on a threonine adjacent to the active site. The identity of the CDK-activating kinase (CAK) that performs this phosphorylation varies across the model organisms. The timing of this phosphorylation varies as well. In mammalian cells, the activating phosphorylation occurs after cyclin binding. In yeast cells, it occurs before cyclin binding. CAK activity is not regulated by known cell-cycle pathways and cyclin binding is the limiting step for CDK activation.

Unlike activating phosphorylation, CDK inhibitory phosphorylation is vital for regulation of the cell cycle. Various kinases and phosphatases regulate their phosphorylation state. One of the kinases that place the tyrosine phosphate is Wee1, a kinase conserved in all eukaryotes. Fission yeast also contains a second kinase Mik1 that can phosphorylate the tyrosine. Vertebrates contain a different second kinase called Myt1 that is related to Wee1 but can phosphorylate both the threonine and the tyrosine. Phosphatases from the Cdc25 family dephosphorylate both the threonine and the tyrosine.

CDK Inhibitors

A cyclin-dependent kinase inhibitor (CKI) is a protein that interacts with a cyclin-CDK complex to block kinase activity, usually during G1 or in response to signals from the environment or from damaged DNA. In animal cells, there are two major CKI families: the INK4 family and the CIP/KIP family. The INK4 family proteins are strictly inhibitory and bind CDK monomers. Crystal structures of CDK6-INK4 complexes show that INK4 binding twists the CDK to distort cyclin binding and kinase activity. The CIP/KIP family proteins bind both the cyclin and the CDK of a complex and can be inhibitory or activating. CIP/KIP family proteins activate cyclin D and CDK4 or CDK6 complexes by enhancing complex formation.

In yeast and Drosophila, CKIs are strong inhibitors of S- and M-CDK, but do not inhibit G1/S-CDKs. During G1, high levels of CKIs prevent cell cycle events from occurring out of order, but do not prevent transition through the Start checkpoint, which is initiated through G1/S-CDKs. Once the cell cycle is initiated, phosphorylation by early G1/S-CDKs leads to destruction of CKIs, relieving inhibition on later cell cycle transitions. In mammalian cells, the CKI regulation works differently. Mammalian protein p27 (Dacapo in Drosophila) inhibits G1/S- and S-CDKs, but does not inhibit S- and M-CDKs.

Suk1 or Cks

The CDKs directly involved in the regulation of the cell cycle associate with small, 9- to 13-kiloDalton proteins called Suk1 or Cks. These proteins are required for CDK function, but their precise role is unknown. Cks1 binds the carboxy lobe of the CDK, and recognizes phosphorylated residues. It may help the cyclin-CDK complex with substrates that have multiple phosphorylation sites by increasing affinity for the substrate.

Non-cyclin CDK Activators

Viral Cyclins

Viruses can encode proteins with sequence homology to cyclins. One much-studied example is K-cyclin (or v-cyclin) from Kaposi sarcoma herpes virus (see Kaposi's sarcoma), which activates CDK6. Viral cyclin-CDK complexes have different substrate specificities and regulation sensitivities.

CDK5 Activators

The proteins p35 and p39 activate CDK5. Although they lack cyclin sequence homology, crystal structures show that p35 folds in a similar way as the cyclins. However, activation of CDK5 does not require activation loop phosphorylation.

RINGO/Speedy

Proteins with no homology to the cyclin family can be direct activators of CDKs. One family of such activators is the RINGO/Speedy family,which was originally discovered in Xenopus. All five members discovered so far directly activate Cdk1 and Cdk2, but the RINGO/Speedy-CDK2 complex recognizes different substrates than cyclin A-CDK2 complex.

Leland H. Hartwell, R. Timothy Hunt, and Paul M. Nurse received the 2001 Nobel Prize in Physiology or Medicine for their complete description of cyclin and cyclin-dependent kinase mechanisms, which are central to the regulation of the cell cycle.

Medical Significance

CDKs are considered a potential target for anti-cancer medication. If it is possible to selectively interrupt the cell cycle regulation in cancer cells by interfering with CDK action, the cell will die. At present, some CDK inhibitors such as seliciclib are undergoing clinical trials. Although it was originally developed as a potential anti-cancer drug, seliciclib has also proven to induce apoptosis in neutrophil granulocytes, which mediate inflammation.This means that novel drugs for treatment of chronic inflammation diseases such as arthritis and cystic fibrosis could be developed.

Flavopiridol (alvocidib) is the first CDK inhibitor to be tested in clinical trials after being identified in an anti-cancer agent screen in 1992. It competes for the ATP site of the CDKs.

More research is required, however, because disruption of the CDK-mediated pathway has potentially serious consequences; while CDK inhibitors seem promising, it has to be determined how side-effects can be limited so that only target cells are affected. As such diseases are currently treated with glucocorticoids, which have often serious side-effects, even a minor success would be an improvement.

Complications of developing a CDK drug include the fact that many CDKs are not involved in the cell cycle, but other processes such as transcription, neural physiology, and glucose homeostasis.

Cyclin-dependent kinase inhibitor drugs

Drug	CDKs Inhibited
Flavopiridol (alvocidib)	1, 2, 4, 6, 7, 9
Olomoucine	1, 2, 5
Roscovitine	1, 2, 5
Purvalanol	1, 2, 5
Paullones	1, 2, 5
Butryolactone	1, 2, 5
Palbociclib	4, 6
Thio/oxoflavopiridols	1
Oxindoles	2
Aminothiazoles	4
Benzocarbazoles	4
Pyrimidines	4
Seliciclib	?

CELL CYCLE CHECKPOINT

Cell cycle checkpoints are control mechanisms in eukaryotic cells which ensure proper division of the cell. Each checkpoint serves as a potential point along the cell cycle, during which the conditions of the cell are assessed, with progression through the various phases of the cell cycle occurring when favorable conditions are met. Currently, there are three known checkpoints: the G1 checkpoint, also known as the restriction or start checkpoint or (Major Checkpoint); the G2/M checkpoint; and the metaphase checkpoint, also known as the spindle checkpoint.

Restriction Point

The **restriction point** (R) is a point in G_1 of the animal cell cycle at which the cell becomes "committed" to the cell cycle and after which extracellular proliferation stimulants are no longer required.

History of the Restriction Point

Originally, Temin showed that chicken cells reach a point at which they are committed to replicate their DNA and are not dependent on extracellular signals. About 20 years later, in 1973, Arthur Pardee demonstrated that a single restriction point exists in G_1. Previously, G_1 had been defined simply as the time between mitosis and S phase. No molecular or morphological place-markers for a cell›s position in G_1 were known. Pardee used a double-block method in which he shifted cells from one cell cycle block (such as critical amino acid withdrawal or serum withdrawal) to another and compared each block's efficiency at preventing progression to S phase. He found that both blocks in all cases examined were equally efficient at blocking S phase progression, indicating that they must all act at the same point in G_1, which he termed the "restriction point," or R-point.

In 1985, Zetterberg and Larsson discovered that, in all stages of the cell cycle, serum deprivation results in inhibition of protein synthesis. Only in postmitotic cells (i.e. cells in early G_1) did serum withdrawal force cells into quiescence (G_0). In fact, Zetterberg found that virtually all of the variability in cell cycle length can be accounted for in the time it takes the cell to move from the restriction point to S phase.

Extracellular Signals and the Restriction Point

Except for early embryonic development, most cells in multicellular organisms persist in a quiescent state known as G_0, where proliferation does not occur, and cells are typically terminally differentiated; other specialized cells continue to divide into adulthood. For both of these groups of cells, a decision has been made to either exit the cell cycle and become quiescent (G_0), or to reenter G_1.

A cell's decision to enter, or reenter, the cell cycle is made before S-phase in G_1 at what is known as the restriction point, and is determined by the combination of promotional and inhibitory extracellular signals that are received and processed. Before the R-point, a cell requires these extracellular stimulants to begin progressing through the first three sub-phases of G_1 (competence, entry G_{1a}, progression G_{1b}). After the R-point has been passed in G_{1b}, however, extracellular signals are no longer required, and the cell is irreversibly committed to preparing for DNA duplication. Further progression is regulated by intracellular mechanisms. Removal of stimulants before the cell reaches the R-point may result in the cell's reversion to quiescence. Under these conditions, cells are actually set back in the cell cycle, and will require additional time (about 8 hours more than the withdrawal time in culture) after passing the restriction point to enter S phase.

Restriction-Point Mechanism

Signals from extracellular growth factors are transduced in a typical manner. Growth factor binds to receptors on the cell surface, and a variety of phosphorylation cascades result in Ca^{2+} uptake and protein phosphorylation. Phosphoprotein levels are counterbalanced by phosphatases. Ultimately, transcriptional activation of certain target genes occurs. Extracellular signaling must be maintained, and the cell must also have access to sufficient nutrient supplies to support rapid protein synthesis. Accumulation of cyclin D›s are essential.

Cyclin D-bound cdk's 4 and 6 are activated by cdk-activating kinase and drive the cell towards the restriction point. Cyclin D, however has a high turnover rate ($t_{1/2}$<25 min). It is because of this quick turnover rate that the cell is extremely sensitive to mitogenic signaling levels, which not only stimulate cycin D production, but also help to stabilize cyclin D within the cell. In this way, cyclin D acts as a mitogenic signal sensor. Cdk inhibitors (CKI), such as the Ink4 proteins and p21, help to prevent improper cyclin-cdk activity.

Active cyclin D-cdk complexes phosphorylate retinoblastoma protein (pRb) in the nucleus. Unphosphorylated Rb acts as an inhibitor of G_1 by preventing E2F-mediated transcription. Once phosphorylated, E2F activates the transcription of cyclins E and A. Active cyclin E-cdk begins to accumulate and completes pRb phosphorylation, as shown in the figure.

Dynamics

A paper published by the Lingchong You and Joe Nevins groups at Duke University in 2008 demonstrated that the a bistable hysteric E2F switch underlies the restriction point. E2F promotes its own activation, and also promotes the inhibition of its own inhibitor (pRb), forming two feedback loops (among others) that are important in establishing bistable systems. The authors of this study used a destabilized GFP-system under the control of the E2F promoter as a readout of E2F activity. Serum-starved cells were stimulated with varying serum concentrations, and the GFP readout was recorded at a single-cell level. They found that the GFP reporter was either on or off, indicating that E2F was either completely activated or deactivated at all of the different serum levels analyzed. Further experiments, in which they analyzed the history-dependence of the E2F system confirmed that it operates as a hysteretic bistable switch.

The Restriction Point in Cancer

Cancer can be seen as a disruption of normal restriction point function, as cells continually and inappropriately reenter the cell cycle, and do not enter G_0. Mutations at many steps in the pathway towards the restriction point can result in cancerous growth of cells. Some of the genes most commonly mutated in cancer include Cdk's and CKI's; overactive Cdk's or underactive CKI's lower the stringency of the restriction point, allowing more cells to bypass senescence.

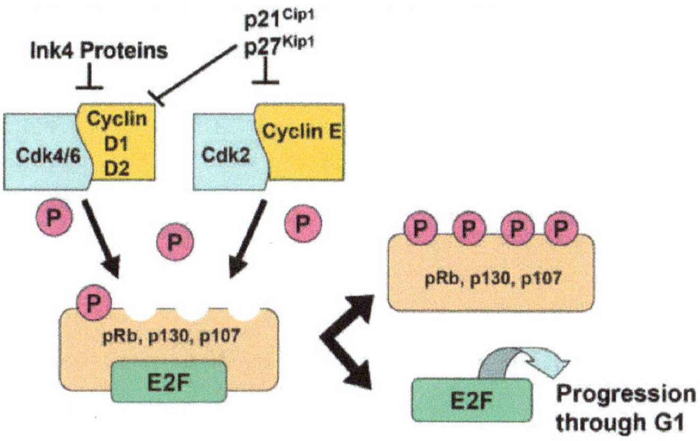

E2F Dynamics at the restriction point

The restriction point is an important consideration in the development of new drug therapies. Under normal physiological conditions, all cell proliferation is regulated by the restriction point. This can be exploited and used as a way to protect non-cancerous cells from chemotherapy treatments. Chemotherapy drugs typically attack cells that are proliferating rapidly. By using drugs that inhibit completion of the restriction point, such as growth factor receptor inhibitors, normal cells are prevented from proliferating, and are thus protected from chemotherapy treatments.

SPINDLE CHECKPOINT

During the process of cell division, the **spindle checkpoint** prevents separation of the duplicated chromosomes until each chromosome is properly attached to the spindle apparatus. In order to preserve the cell›s identity and proper function, it is necessary to maintain the appropriate number of chromosomes after each cell division. An error in generating daughter cells with fewer or greater number of chromosomes than expected (a situation termed aneuploidy), may lead in best case to cell death, or alternatively it may generate catastrophic phenotypic results. Examples include:

- In cancer cells, aneuploidy is a frequent event, indicating that these cells present a defect in the machinery involved in chromosome segregation, as well as in the mechanism ensuring that segregation is correctly performed.
- In humans, Down syndrome appears in children carrying in their cells one extra copy of chromosome 21, as a result of a defect in chromosome segregation during meiosis in one of the progenitors. This defect will generate a gamete (spermatozoide or oocyte) with an extra chromosome 21. After fertilisation, this gamete will generate an embryo with three copies of chromosome 21.

The mechanisms verifying that all the requirements to pass to the next phase in the cell cycle have been fulfilled are called checkpoints. All along the cell cycle, there are different checkpoints. The checkpoint ensuring that chromosome segregation is correct is termed **spindle assembly checkpoint** (SAC), spindle checkpoint or mitotic checkpoint.

During mitosis or meiosis, the spindle checkpoint prevents anaphase onset until all chromosomes are properly attached to the spindle. To achieve proper segregation, the two kinetochores on the sister chromatids must be attached to opposite spindle poles (bipolar orientation). Only this pattern of attachment will ensure that each daughter cell receives one copy of the chromosome.

Chromosome Segregation

Cell Division: Duplication of Material and Distribution to Daughter Cells

When cells are ready to divide, because cell size is big enough or because they receive the appropriate stimulus, they activate the mechanism to enter into the cell cycle, and they duplicate most organelles during S (synthesis) phase, including their centrosome. Therefore, when the cell division process will end, each daughter cell will receive a complete set of organelles. At the same time, during S phase all cells must duplicate their DNA very precisely, a process termed DNA replication. Once DNA replication has finished, in eukaryotes the DNA molecule is compacted and condensed, to form the mitotic chromosomes, each one constituted by two sister chromatids, which stay held together by the establishment of cohesion between them; each chromatid is a complete DNA molecule, attached via microtubules to one of the two centrosomes of the dividing cell, located at opposed poles of the cell. The structure formed by the centrosomes and the microtubules is named mitotic spindle, due to its characteristic shape, holding the chromosomes between the two centrosomes. Both sister chromatids stay together until anaphase; at this moment they separate from each other and they travel towards the centrosome to which they are attached. In this way, when the two daughter cells separate at the end of the division process, each one will receive a complete set of chromatids. The mechanism responsible for the correct distribution of sister chromatids during cell division is named **chromosome segregation**.

To ensure that chromosome segregation takes place correctly, cells have developed a precise and complex mechanism. In the first place, cells must coordinate centrosome duplication with DNA replication, and a failure in this coordination will generate monopolar or multipolar mitotic spindles, which generally will produce abnormal chromosome segregation, because in this case, chromosome distribution will not take place in a balanced way.

Mitosis: Anchoring of Chromosomes to the Spindle and Chromosome Segregation

During S phase, the centrosome starts to duplicate. Just at the beginning of mitosis, both centrioles achieve their maximal length, recruit additional material and their capacity to nucleate microtubules increases. As mitosis progresses, both centrosomes separate to generate the mitotic spindle. In this way, the mitotic spindle has two poles emanating microtubules. Microtubules (MTs) are long proteic filaments, with asymmetric extremities:

one end termed "minus" (-) end, relatively stable and close to the centrosome, and an end termed "plus" (+) end, with alternating phases of growth and retraction, exploring the center of the cell searching the chromosomes. Each chromatid has a special region, named the centromere, on top of which is assembled a proteic structure termed kinetochore, which is able to stabilize the microtubule plus end.

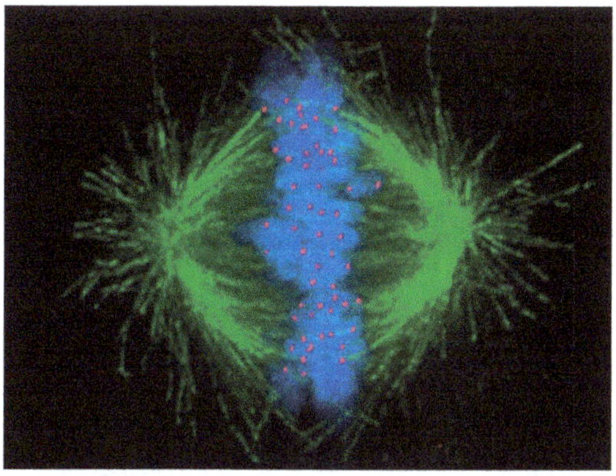

Image of a human cell during mitosis; microtubules are shown in green (forming the mitotic spindle), chromosomes are in blue in the spindle equator and kinetochores in red.

Therefore, if by chance a microtubule exploring the center of the cell encounters a kinetochore, it may happen that the kinetochore will capture it, so that the chromosome will become attached to the spindle via the kinetochore of one of its sister chromatids. The chromosome plays an active role in the attachment of kinetochores to the spindle. Bound to the chromatin is a Ran guanine nucleotide exchange factor (GEF) that stimulates cytosolic Ran near the chromosome to bind GTP in place of GDP. The activated GTP-bound form of Ran releases microtubule-stabilizing proteins, such as TPX2, from protein complexes in the cytosol, which induces nucleation and polymerization of microtubules around the chromosomes. These kinetochore-derived microtubules, along with kinesin motor proteins in the outer kinetochore, facilitate interactions with the lateral surface of a spindle pole-derived microtubule. These lateral attachments are unstable, however, and must be converted to an end-on attachment. Conversion from lateral to end-on attachments allows the growth and shrinkage of the microtubule plus-ends to be converted into forces that push and pull chromosomes to achieve proper bi-orientation. As it happens that sister chromatids are attached together and both kinetochores are located back-to-back on both chromatids, when one kinetochore becomes attached to one centrosome, the sister kinetochore becomes exposed to the centrosome located in the opposed pole; for this reason, in most cases the second kinetochore becomes associated to the centrosome in the opposed pole, via its microtubules, so that the chromosomes become bi-oriented, a fundamental configuration (also named *amphitelic*) to ensure that chromosome segregation will take place correctly when the cell will divide. Occasionally, one of the two sister kinetochores may attach simultaneously to MTs generated by both poles, a configuration named *merotelic*, which is not detected by the spindle checkpoint but that may generate lagging chromosomes during anaphase and, consequently, aneuploidy. Merotelic orientation (characterized by the absence of tension between sister kinetochores) is frequent at the beginning of mitosis, but the protein Aurora B (a kinase conserved from yeast to vertebrates) detects and eliminates this type of anchoring. (Note: Aurora B is frequently overexpressed in various types of tumors and currently is a target for the development of anticancer drugs.)

Discovery of the spindle assembly checkpoint (SAC)

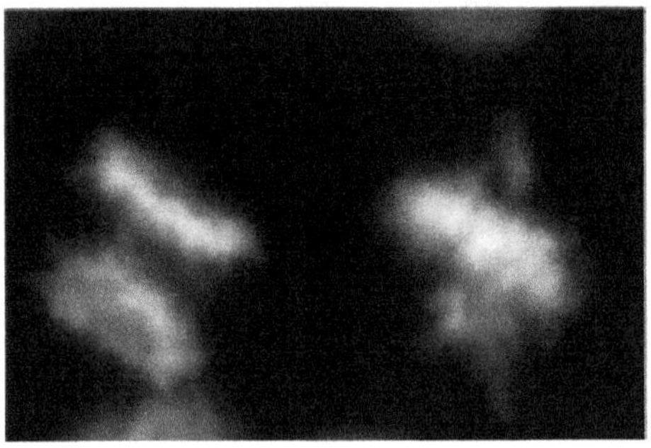

Microscopy image showing two cells with their chromosomes stained with DAPI, one at anaphase (left) and the other in metaphase (right), with most of its chromosomes in the metaphase plate and some chromosomes still not aligned.

Zirkle (in 1970) was one of the first researchers to observe that, when just one chromosome is retarded to arrive at the metaphase plate, anaphase onset is postponed until some minutes after its arrival. This observation, together with similar ones, suggested that a control mechanism exists at the metaphase-to-anaphase transition. Using drugs such as nocodazole and colchicine, the mitotic spindle disassembles and the cell cycle is blocked at the metaphase-to-anaphase transition. Using these drugs (see the review from Rieder and Palazzo in 1992), the putative control mechanism was named **Spindle Assembly Checkpoint** (SAC). This regulatory mechanism has been intensively studied since(see the review from Burke and Stukenberg in 2008).

Using different types of genetic studies, it has been established that diverse kinds of defects are able to activate the SAC: spindle depolimerization, the presence of dicentric chromosomes (with two centromeres), centromeres segregating in an aberrant way, defects in the spindle pole bodies in *S. cerevisiae*, defects in the kinetochore proteins, mutations in the centromeric DNA or defects in the molecular motors active during mitosis. A summary of these observations can be found in the article from Hardwick and collaborators in 1999.

Using its own observations, Zirkle was the first to propose that some (...) susbstance, necessary for the cell to proceed to anaphase, appears some minutes after C (moment of the arrival of the last chromosome to the metaphase plate), or after a drastic change in the cytoplasmic condition, just at C or immediately after C", suggesting that this function is located on kinetochores unattached to the mitotic spindle. McIntosh extended this proposal, suggesting that one enzyme sensitive to tension located at the centromeres produces an inhibitor to the anaphase onset when the two sister kinetochores are not under bipolar tension. Indeed, the available data suggested that the signal wait to enter in anaphase is produced mostly on or close to unattached kinetochores. However, the primary event associated to the kinetochore attachment to the spindle, which is able to inactivate the inhibitory signal and release the metaphase arrest, could be either the acquisition of microtubules by the kinetochore (as proposed by Rieder and collaborators in 1995), or the tension stabilizing the anchoring of microtubules to the kinetochores (as suggested by the experiments realized at Nicklas› lab). Subsequent studies in cells containing two independent mitotic spindles in a sole cytoplasm showed that the inhibitor of the metaphase-to-anaphase transition is generated by unattached kinetochores and is not freely diffusible in the cytoplasm.Yet in the same study it was shown that, once the transition from metaphase to anaphase is initiated in one part of the cell, this information is extended all along the cytoplasm, and can overcome the signal wait to enter in anaphase associated to a second spindle containing unattached kinetochores.

Sister Chromatids Cohesion During Mitosis

Cohesin: SMC Proteins

As it has been previously noted, sister chromatids stay associated from S phase (when DNA is replicated to generate two identic copies, the two chromatids) until anaphase. At this point, the two sister chromatids separate and travel to opposite poles in the dividing cell. Genetic and biochemical studies in yeast and in egg's extracts in *Xenopus laevis* identified a polyprotein complex as an essential player in sister chromatids cohesion (see the review from Hirano in 2000). This complex is known as the cohesin complex and in *Saccharomyces cerevisiae* is composed of at least four subunits: Smc1p, Smc3p, Scc1p (or Mcd1p) and Scc3p. Both Smc1p and Smc3p belong to the family of proteins for the **Structural Maintenance of Chromosomes** (SMC), which constitute a group of chromosomic ATPases highly conserved, and form an heterodimer (Smc1p/Smc3p). Scc1p is the homolog in *S.cerevisiae* of Rad21, first identified as a protein involved in DNA repair in *S. pombe*. These four proteins are essential in yeast, and a mutation in any of them will produce premature sister chromatid separation. In yeast, cohesin binds to preferential sites along chromosome arms, and is very abundant close to the centromeres, as it was shown in a study using chromatin immunoprecipitation.

The Role of Heterochromatin

Classical cytologic observations suggested that sister chromatids are more strongly attached at heterochromatic regions, and this suggested that the special structure or composition of heterochromatin might favour cohesin recruitment. In fact, it has been shown that Swi6 (the homolog of HP-1 in *S. pombe*) binds to methylated Lys 9 of histone H3 and promotes the binding of cohesin to the centromeric repeats in *S. pombe*.More recent studies indicate that the RNAi machinery regulates heterochromatin establishment, which in turn recruits cohesin to this region, both in *S. pombe* and in vertebrate cells. However, there must be other mechanisms than heterochromatin to ensure an augmented cohesion at centromeres, because *S. cerevisiae* lacks heterochromatin next to centromeres, but the presence of a functional centromere induces an increase of cohesin association in a contiguous region, spanning 20-50kb.

In this direction, **Orc2** (one protein included in the origin recognition complex, ORC, implicated in the initiation of DNA replication during S phase) is also located on kinetochores during mitosis in human cells; in agreement with this localization, some observations indicate that Orc2 in yeast is implicated in sister chromatid cohesion, and its removal induces SAC activation. It has also been observed that other components of the ORC complex (such as orc5 in *S. pombe*) are implicated in cohesion. However, the molecular pathway involving the ORC proteins seems to be additive to the cohesins' pathway, and it is mostly unknown.

Function of Cohesion and its Dissolution

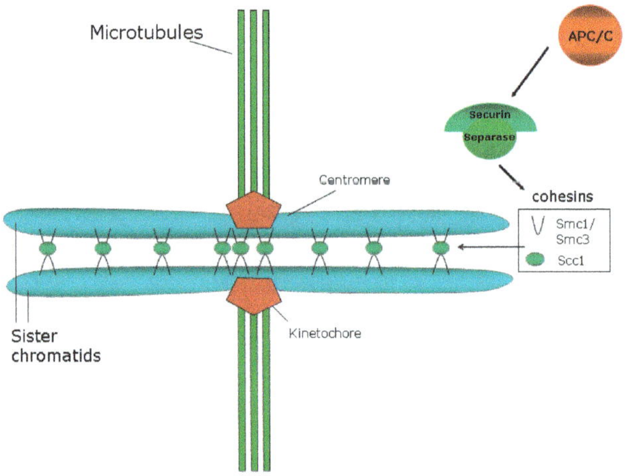

Scheme showing sister chromatids cohesion, anchored to spindle microtubules via their kinetochores.

Centromeric cohesion resists the forces exerted by spindle microtubules towards the poles, which generate tension between sister kinetochores. In turn, this tension stabilizes the attachment microtubule-kinetochore, through a mechanism implicating the protein Aurora B (a review about this issue : Hauf and Watanabe 2004).

Indeed, a decrease in the cellular levels of cohesin generates the premature separation of sister chromatids, as well as defects in chromosome congression at the metaphase plate and delocalization of the proteins in the **chromosomal passenger complex**, which contains the protein Aurora B. The proposed structure for the cohesin complex suggests that this complex connects directly both sister chromatids. In this proposed structure, the SMC components of cohesin play a structural role, so that the SMC heterodimer may function as a DNA binding protein, whose conformation is regulated by ATP. Scc1p and Scc3p, however, would play a regulatory role.

In *S. cerevisiae*, Pds1p (also known as securin) regulates sister chromatids cohesion, because it binds and inhibits the protease Esp1p (**separin** or **separase**). When anaphase onset is triggered, the anaphase-promoting complex (**APC/C** or Cyclosome) degrades securin. APC/C is a ring E3 ubiquitin ligase that recruits an E2 ubiquitin-conjugating enzyme loaded with ubiquitin. Securin is recognized only if Cdc20, the activator subunit, is bound to the APC/C core. When securin, Cdc20, and E2 are all bound to APC/C E2 ubiquitinates securin and selectively degrades it. Securin degradation releases the protease Esp1p/separase, which degrades the cohesin rings that link the two sister chromatids, therefore promoting sister chromatids separation. It has been also shown that Polo/Cdc5 kinase phosphorylates serine residues next to the cutting site for Scc1, and this phosphorylation would facilitate the cutting activity.

Although this machinery is conserved through evolution, in vertebrates most cohesin molecules are released in prophase, independently of the presence of the APC/C, in a process dependent on Polo-like 1 (PLK1) and Aurora B. Yet it has been shown that a small quantity of Scc1 remains associated to centromeres in human cells until metaphase, and a similar amount is cut in anaphase, when it disappears from centromeres. On the other hand, some experiments show that sister chromatids cohesion in the arms is lost gradually after sister centromeres have separated, and sister chromatids move toward the opposite poles of the cell.

According to some observations, a fraction of cohesins in the chromosomal arms and the centromeric cohesins are protected by the protein **Shugoshin** (Sgo1), avoiding their release during prophase. To be able to function as protector for the centromeric cohesion, Sgo1 must be inactivated at the beginning of anaphase, as well as Pds1p. In fact, both Pds1p and Sgo1 are substrates of APC/C in vertebrates.

Metaphase to Anaphase Transition

The beginning of metaphase is characterized by the connection of the microtubules to the kinetochores of the chromosomes, as well as the alignment of the chromosomes in the middle of the cell. Each chromatid has its own kinetochore, and all of the microtubules that are bound to kinetochores of sister chromatids radiate from opposite poles of the cell. These microtubules exert a pulling force on the chromosomes towards the opposite ends of the cells, while the cohesion between the sister chromatids opposes this force.

At the metaphase to anaphase transition, this cohesion between sister chromatids is dissolved, and the separated chromatids are pulled to opposite sides of the cell by the spindle microtubules. The chromatids are further separated by the physical movement of the spindle poles themselves. Premature dissociation of the chromatids can lead to chromosome missegregation and aneuploidy in the daughter cells. Thus, the job of the metaphase checkpoint is to prevent this transition into anaphase until the chromosomes are properly attached, before the sister chromatids separate.

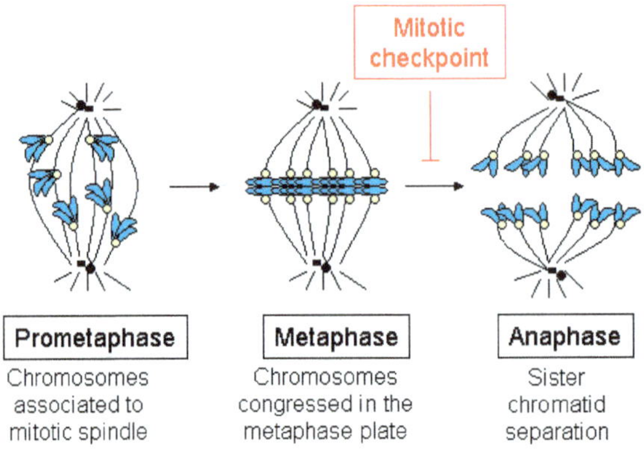

Scheme showing cell cycle progression between prometaphase and anaphase.

Spindle Assembly Checkpoint Overview

The spindle assembly checkpoint (SAC) is an active signal produced by improperly attached kinetochores, which is conserved in all eukaryotes. The SAC stops the cell cycle by negatively regulating CDC20, thereby preventing the activation of the polyubiquitylation activities of anaphase promoting complex (APC). The proteins responsible for the SAC signal compose the mitotic checkpoint complex (MCC), which includes SAC proteins, MAD2/MAD3 (mitotic arrest deficient), BUB3 (budding uninhibited by benzimidazole), and CDC20. Other proteins involved in the SAC include MAD1, BUB1, MPS1, and Aurora B. For higher eukaryotes, additional regulators of the SAC include constituents of the ROD-ZW10 complex, p31^{comet}, MAPK, CDK1-cyclin-B, NEK2, and PLK1.

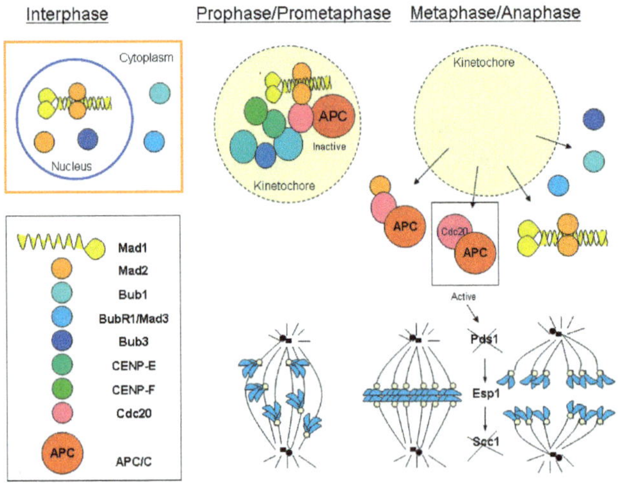

Checkpoint Activation

The SAC monitors the interaction between improperly connected kinetochores and spindle microtubules, and is maintained until kinetochores are properly attached to the spindle. During prometaphase, CDC20 and the SAC proteins concentrate at the kinetochores before attachment to the spindle assembly. These proteins keep the SAC activated until they are removed and the correct kinetochore-microtubule attachment is made. Even a single unattached kinetochore can maintain the spindle checkpoint. After attachment of microtubule plus-ends and formation of kinetochore microtubules, MAD1 and MAD2 are depleted from the kinetochore assembly. Another regulator of checkpoint activation is kinetochore tension. When sister kinetochores are properly attached to opposite spindle poles, forces in the mitotic spindle generate tension at the kinetochores. Bi-oriented sister kinetochores stabilize the kinetochore-microtubule assembly whereas weak tension has a destabilizing effect. In response to incorrect kinetochore attachments such as syntelic attachment, where both kinetochores becomes attached to one spindle pole, the weak tension generated destabilizes the incorrect attachment and allows the kinetochore to reattach correctly to the spindle body. During this process, kinetochores that are attached to the mitotic spindle but that are not under tension trigger the spindle checkpoint. Aurora-B/Ipl1 kinase of the chromosomal passenger complex functions as the tensions sensor in improper kinetochore attachments. It detects and destabilizes incorrect attachments through control of the microtubule-severing KINI kinesin MCAK, the DASH complex, and the Ndc80/Hec1 complex at the microtubule-kinetochore interface. The Aurora-B/Ipl1 kinase is also critical in correcting merotelic attachments, where one kinetochore is simultaneously attached to both spindle poles. Merotelic attachments generate sufficient tension and are not detected by the SAC, and without correction, may result in chromosome mis-segregation due to slow chromatid migration speed. While microtubule attachment is independently required for SAC activation, it is unclear whether tension is an independent regulator of SAC, although it is clear that differing regulatory behaviors arise with tension.

Once activated, the spindle checkpoint blocks anaphase entry by inhibiting the anaphase-promoting complex via regulation of the activity of mitotic checkpoint complex. The mechanism of inhibition of APC by the mitotic checkpoint complex is poorly understood, although it is hypothesized that the MCC binds to APC as a pseudosubstrate using the KEN-box motif in BUBR1. At the same time that mitotic checkpoint complex is being activated, the centromere protein CENP-E activates BUBR1, which also blocks anaphase.

Mitotic Checkpoint Complex formation

The mitotic checkpoint complex is composed of BUB3 together with MAD2 and MAD3 bound to Cdc20. MAD2 and MAD3 have distinct binding sites on CDC20, and act synergistically to inhibit APC/C. The MAD3 complex is composed of BUB3, which binds to Mad3 and BUB1B through the short linear motif known as the GLEBS motif. The exact order of attachments which must take place in order to form the MCC remains unknown. It is possible that Mad2-Cdc20 form a complex at the same time as BUBR1-BUB3-Cdc20 form another complex, and these two subcomplexes are consequently combined to form the mitotic checkpoint complex. In human cells, binding of BUBR1 to CDC20 requires prior binding of MAD2 to CDC20, so it is possible that the MAD2-CDC20 subcomplex acts as an initiator for MCC formation. BUBR1 depletion leads only to a mild reduction in Mad2-Cdc20 levels while Mad2 is required for the binding of BubR1-Bub3 to Cdc20. Nevertheless BUBR1 is still required for checkpoint activation.

The mechanism of formation for the MCC is unclear and there are competing theories for both kinetochore-dependent and kinetochore-independent formation. In support of the kinetochore-independent theory, MCC is detectable in *S. cerevisiae* cells in which core kinetocore assembly proteins have been mutated and cells in which the SAC has been deactivated, which suggests that the MCC could be assembled during mitosis without kinetochore localization. In one model, unattached prometaphase kinetochores can 'sensitize' APC to inhibition of MCC by recruiting the APC to kinetochores via a functioning SAC. Furthermore, depletions of various SAC proteins have revealed that MAD2 and BUBR1 depletions affect the timing of mitosis independently of kinetochores, while depletions of other SAC proteins result in a dysfunctional SAC without altering the duration of mitosis. Thus it is possible that the SAC functions through a two-stage timer where MAD2 and BUBR1 control the duration of mitosis in the first stage, which may be extended in the second stage if there are unattached kinetochores as well as other SAC proteins. However, there are lines of evidence which are in disfavor of the kinetochore-independent assembly. MCC has yet to be found during interphase, while MCC does not form from its constituents in *X. laevis* meiosis II extracts without the addition of sperm of nuclei and nocodazole to prevent spindle assembly.

The leading model of MCC formation is the MAD2-template model, which depends on the kinetochore dynamics of MAD2 to create the MCC. MAD1 localizes to unattached kinetochores while binding strongly to MAD2. The localization of MAD2 and BubR1 to the kinetochore may also be dependent on the Aurora B kinase. Cells lacking Aurora B fail to arrest in metaphase even when chromosomes lack microtubule attachment.Unattached kinetochores first bind to a MAD1-C-MAD2-p31^{comet} complex and releases the p31^{comet} through unknown mechanisms. The resulting MAD-C-MAD2 complex recruits the open conformer of Mad2 (O-Mad2) to the kinetochores. This O-Mad2 changes its conformation to closed Mad2 (C-Mad2) and binds Mad1. This Mad1/C-Mad2 complex is responsible for the recruitment of more O-Mad2 to the kinetochores, which changes its conformation to C-Mad2 and binds Cdc20 in an auto-amplification reaction. Since MAD1 and CDC20 both contain a similar MAD2-binding motif, the empty O-MAD2 conformation changes to C-MAD2 while binding to CDC20. This positive feedback loop is negatively regulated by p31^{comet}, which competitively binds to C-MAD2 bound to either MAD1 or CDC20 and reduces further O-MAD2 binding to C-MAD2. Further control mechanisms may also exist, considering that p31^{comet} is not present in lower eukaryotes. The 'template model' nomenclature is thus derived from the process where MAD1-C-MAD2 acts as a template for the formation of C-MAD2-CDC20 copies. This sequestration of Cdc20 is essential for maintaining the spindle checkpoint.

Checkpoint Deactivation

Several mechanisms exist to deactivate the SAC after correct bi-orientation of sister chromatids. Upon microtubule-kinetochore attachment, a mechanism of stripping via a dynein-dynein motor complex transports spindle checkpoint proteins away from the kinetochores. The stripped proteins, which include MAD1, MAD2, MPS1, and CENP-F, are then redistributed to the spindle poles. The stripping process is highly dependent on undamaged microtubule structure as well as dynein motility along microtubules. As well as functioning as a regulator of the C-MAD2 positive feedback loop, p31^{comet} also may act as a deactivator of the SAC. Unattached kinetochores temporarily inactivate p31, but attachment reactivates the protein and inhibits MAD2 activation, possibly by inhibitory phosphorylation. Another possible mechanism of SAC inactivation results from energy-dependent dissociation of the MAD2-CDC20 complex through non-degradative ubiquitylation of CDC20. Conversely, the de-ubiquitylating enzyme protectin is required to maintain the SAC. Thus, unattached kinetochores maintain the checkpoint by continuously recreating the MAD2-CDC20 subcomplex from its components. The SAC may also be deactivated by APC activation induced proteolysis. Since the SAC is not reactivated by the loss of sister-chromatid cohesion during anaphase, the proteolysis of cyclin B and inactivation of the CDK1-cyclin-B kinase also inhibits SAC activity. Degradation of MPS1 during anaphase prevents the reactivation of SAC after removal of sister-chromatid cohesion. After checkpoint deactivation and during the normal anaphase of the cell cycle, the anaphase promoting complex is activated through decreasing MCC activity. When this happens the enzyme complex polyubiquitinates the anaphase inhibitor securin. The ubiquitination and destruction of securin at the end of metaphase releases the active protease called separase. Separase cleaves the cohesion molecules that hold the sister chromatids together to activate anaphase.

New Model for SAC Deactivation in S. cerevisiae: The Mechanical Switch

A new mechanism has been suggested to explain how end-on microtubule attachment at the kinetochore is able to disrupt specific steps in SAC signaling. In an unattached kinetochore, the first step in the formation of the MCC is phosphorylation of Spc105 by the kinase Mps1. Phosphorylated Spc105 is then able to recruit the downstream signaling proteins Bub1 and 3; Mad 1,2, and 3; and Cdc20. Association with Mad1 at unattached kinetochores causes Mad2 to undergo a conformational change that converts it from an open from (O-Mad2) to a closed form (C-Mad2.) The C-Mad2 bound to Mad1 then dimerizes with a second O-Mad2 and catalyzes its closure around Cdc20. This C-Mad2 and Cdc20 complex, the MCC, leaves Mad1 and C-Mad2 at the kinetochore to form another MCC. The MCCs each sequester two Cdc20 molecules to prevent their interaction with the APC/C, thereby maintaining the SAC. Mps1's phosphorylation of Spc105 is both necessary and sufficient to initiate the SAC signaling pathway, but this step can only occur in the absence of microtubule attachment to the kinetochore. Endogenous Mps1 is shown to associate with the calponin-homology (CH) domain of Ndc80, which is located in the outer kinetochore region that is distant from the chromosome. Though Mps1 is docked in the outer kinetochore, it is still able to localize within the inner kinetochore and phosphorylate Spc105 because of flexible hinge regions on Ndc80. However, the mechanical switch model proposes that end-on attachment of a microtubule to the kinetochore deactivates the SAC through two mechanisms. The presence of an attached microtubule increases the distance between the Ndc80 CH domain and Spc105. Additionally, Dam1/DASH, a large complex consisting of 160 proteins that forms a ring around the attached microtubule, acts as a barrier between the two proteins. Separation prevents interactions between Mps1 and Spc105 and thus inhibits the SAC signaling pathway.

It is important to note that this model is not applicable to SAC regulation in higher order organisms, including animals. A main facet of the mechanical switch mechanism is that in *S. cerevisiae* the structure of the kinetochore only allows for attachment of one microtubule. Kinetochores in animals, on the other hand, are much more complex meshworks that contain binding sites for a multitude of microtubules. Microtubule attachment at all of the kinetochore binding sites is not necessary for deactivation of the SAC and progression to anaphase. Therefore, microtubule-attached and microtubule-unattached states coexist in the animal kinetochore while the SAC is inhibited. This model does not include a barrier that would prevent Mps1 associated with an attached kinetochore from phosphorylating Spc105 in an adjacent unattached kinetochore. Furthermore, the yeast Dam1/DASH complex is not present in animal cells.

Spindle Checkpoint Defects and Cancer

When the spindle checkpoint misfunctions, this can lead to chromosome missegregation, aneuploidy and even tumorigenesis. Transformation occurs and is accelerated when maintenance of genomic integrity breaks down especially at the gross level of whole chromosomes or large portions of them. In fact, aneuploidy is the most common characteristic of human solid tumors and thus the spindle assembly checkpoint might be regarded as a possible target for anti-tumour therapy. This is a much underappreciated fact since mutations in specific genes known as oncogenes or tumor suppressor are primarily thought to be behind genetic instability and tumorigenesis. Usually the various checkpoints in the cell cycle take care of genomic integrity via highly conserved redundant mechanisms that are important for maintaining cellular homeostasis and preventing tumorigenesis. Several spindle assembly checkpoint proteins act both as positive and negative regulators to ensure the proper chromosome segregation in each cell cycle preventing chromosome instability (CIN) also known as genome instability.

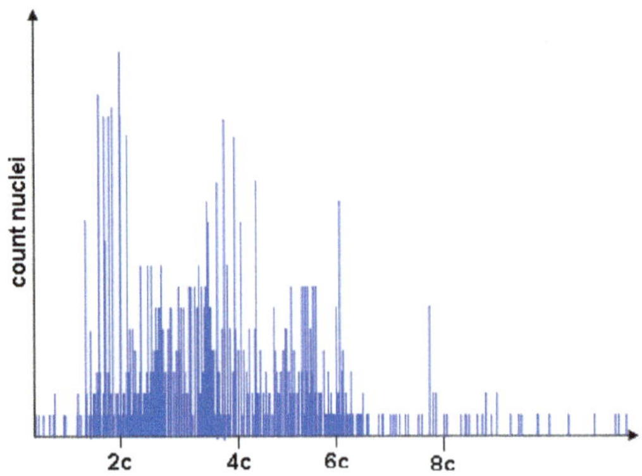

Cytometric analysis of malignant carcinoma displaying aneuploidy.

Genomic integrity is now appreciated at several levels where some tumors display instability manifested as base substitutions, insertions, and deletions, while the majority displays gains or losses of whole chromosomes.

Due to the fact that alterations in mitotic regulatory proteins can lead to aneuploidy and this is a frequent event in cancer, it was initially thought that these genes could be mutated in cancerous tissues.

Mutated Genes in Cancers

In some cancers the genes that underlie the defects resulting in transformation are well characterized. In the hematological cancers such as multiple myeloma cytogenetic abnormalities are very common due to the inherent nature of DNA breaks needed for immunoglobulin gene rearrangement. However, defects in proteins such as MAD2 that function predominantly at the SAC also are characterized in multiple myeloma. Most solid tumors are also predominantly aneuploid. For colorectal cancer, BUB1 and BUBR1 and amplification of STK15 are key regulators that have been implicated in the genomic instability resulting in cancer. In breast cancer, the genetic form characterized by the BRCA-1 gene exhibits greater levels of genomic instability than sporadic forms. Experiments showed that BRCA-1 null mice have decreased expression of the key spindle checkpoint protein MAD2 . For other cancers, more work is warranted to identify the causes of aneuploidy.

Other Genes not Traditionally Associated with the SAC in Cancer

Clearly variations in the physiological levels of these proteins (such as Mad2 or BubR1) are associated with aneuploidy and tumorigenesis, and this has been demonstrated using animal models. However, recent studies indicate that what seems to happen is a more complicated scenario: aneuploidy would drive a high incidence of tumorigenesis only when alterations in the levels of specific mitotic checkpoint components (either reduction or overexpression) in tissues is also inducing other defects able to predispose them to tumors. That is, defects such as an increase in DNA damage, chromosomal rearrangements, and/or a decreased incidence of cell death. For some mitotic checkpoint components, it is known that they are implicated in functions outside mitosis: nuclear import (Mad1), transcriptional repression (Bub3), and cell death, DNA damage response, aging, and megakaryopoiesis for BubR1. All this supports the conclusion that increase in tumorigenesis is associated with defects other than aneuploidy alone.

Cancer-associated mutations affecting known checkpoint genes like BUB1 or BUBR1 are actually rare. However, several proteins implicated in cancer have intersections to spindle assembly networks. Key tumor suppressors such as p53 also play a role in the spindle checkpoint. Absence of p53, the most commonly mutated gene in human cancer, has a major effect on cell cycle checkpoint regulators and has been shown to act at the G1 checkpoint in the past, but now appears to be important in regulating the spindle checkpoint as well. Another key aspect of cancer is inhibition of cell death or apoptosis. Survivin, a member of the inhibitor of apoptosis (IAP) family, is localized in pools at microtubules of the mitotic spindle near the centrosomes and at the kinetochores of metaphase chromosomes. Not only does survivin inhibit apoptosis to promote tumorigenesis, but it has been implicated (through experimental knockout mice) as an important regulator of chromosome segregation, and late stage mitosis similar to its role in more primitive organisms.

Other aspects of the spindle assembly checkpoint such as kinetochore attachment, microtubule function, and sister chromatid cohesion are likely to be defective as well to cause aneuploidy. Cancer cells have been observed to divide in multiple directions by evading the spindle assembly checkpoint resulting in multipolar mitoses. The multipolar metaphase-anaphase transition occurs through an incomplete separase cycle that results in frequent nondisjunction events which amplify aneuploidy in cancer cells.

SAC cancer Therapies

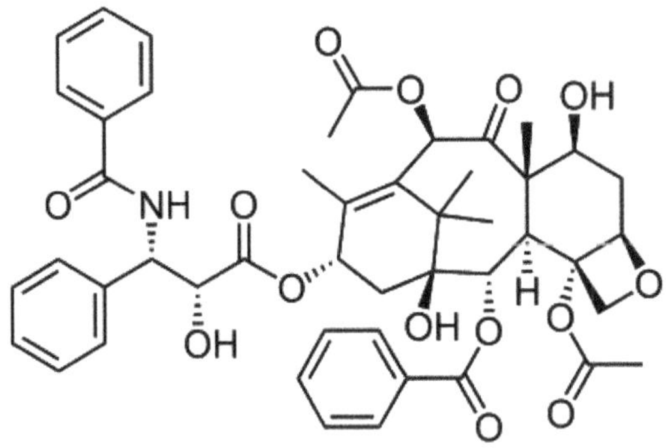

The chemical structure of paclitaxel or TAXOL, a mitotic inhibitor used in cancer chemotherapy.

Advances in this field have led to the introduction of development of some therapies targeted at spindle assembly defects. Older treatments such as vinca alkaloids and taxanes target microtubules that accompany mitotic spindle formation via disruption of microtubule dynamics which engage the SAC arresting the cell and eventually leading to its death. taxol and Docetaxel both are still used in the treatment of breast cancer, ovarian cancer and other types of epithelial cancer. However, these treatments are often characterized by high rates of side effects and drug resistance.

Other targets within the network of regulators that influence the SAC are also being pursued; strong interest has shifted towards the aurora kinase proteins. The kinase gene Aurora A when amplified acts as an oncogene overriding the SAC leading to abnormal initiation of anaphase and subsequent aneuploidy and also resistance to TAXOL . Excitingly, a small molecule inhibitor of Aurora A has shown antitumor effects in an in vivo model suggesting that this might be a good target for further clinical development. Aurora B inhibitors, which are also in clinical development lead to abnormal kinetochore to microtubule attachment and abrogate the mitotic checkpoint as well. Survivin is also an attractive molecular target for clinical therapeutic development as it acts as a major node in a multitude of pathways, one of which is spindle formation and checkpoint control. Even further approaches have included a look at inhibition of mitotic motor proteins like KSP. These inhibitors, which have recently entered clinical trials, cause mitotic arrest and by engaging the spindle assembly checkpoint and induce apoptosis.

REFERENCES

1. Carter, J. Stein (2014-01-14). Mitosis. biology.clc.uc.edu.
2. Cell Division: Stages of Mitosis | Learn Science at Scitable. www.nature.com. Retrieved 2015-11-16.
3. Maton, A.; Hopkins, J. J.; LaHart, S; Quon Warner, D.; Wright, M.; Jill, D. (1997). Cells: Building Blocks of Life. New Jersey: Prentice Hall. pp. 70–4. ISBN 0-13-423476-6.
4. Kalatova, B.; Jesenska, R.; Hlinka, D.; Dudas, M. (2015). Tripolar mitosis in human cells and embryos: Occurrence, pathophysiology and medical implications. Acta Histochemica. **117** *(1): 111–125. doi:10.1016/j.acthis.2014.11.009. PMID 25554607.*
5. Kops, G. J.; Weaver, B. A.; Cleveland, D. W. (October 2005). On the road to cancer: Aneuploidy and the mitotic checkpoint. Nature Reviews Cancer. **5** *(10): 773–785. doi:10.1038/nrc1714. PMID 16195750.*
6. Raikov, IB (1994). The diversity of forms of mitosis in protozoa: A comparative review. European Journal of Protistology. **30** *(3): 253–69. doi:10.1016/S0932-4739(11)80072-6.*
7. De Souza, C. P.; Osmani, S. A. (2007). Mitosis, Not Just Open or Closed. Eukaryotic Cell. **6** *(9): 1521–7. doi:10.1128/EC.00178-07. PMC 2043359. PMID 17660363.*
8. Fokin, S. I. (2013). Otto Bütschli (1848–1920) Where we will genuflect? (PDF). Protistology. **8** *(1): 22–35.*
9. Sharp, L. W. (1921). Introduction To Cytology. New York: McGraw Hill Book Company Inc. p. 143.
10. Janusz Komender (2008). Kilka słów o doktorze Wacławie Mayzlu i jego odkryciu [On Waclaw Mayzel and his observation of mitotic division] (PDF). Postępy Biologii Komórki (in Polish). Polskie Towarzystwo Anatomiczne, Polskie Towarzystwo Biologii Komórki. **35** *(3): 405–407.*
11. Iłowiecki, Maciej (1981). Dzieje nauki polskiej. Warszawa: Wydawnictwo Interpress. p. 187. ISBN 83-223-1876-6.
12. mitosis. Online Etymology Dictionary.
13. μίτος. Liddell, Henry George; Scott, Robert; A Greek–English Lexicon at the Perseus Project
14. Blow, J.; Tanaka, T. (2005). The chromosome cycle: coordinating replication and segregation: Second in the cycles review series. EMBO Reports. **6** *(11): 1028–34. doi:10.1038/sj.embor.7400557. PMC 1371039. PMID 16264427.*

15. Zhou, J.; Yao, J.; Joshi, H. (2002). Attachment and tension in the spindle assembly checkpoint. Journal of Cell Science. **115** *(Pt 18): 3547–55. doi:10.1242/jcs.00029. PMID 12186941.*

16. Lloyd, C.; Chan, J. (2006). Not so divided: the common basis of plant and animal cell division. Nature reviews. Molecular cell biology. **7** *(2): 147–52. doi:10.1038/nrm1831. PMID 16493420.*

17. Raven et al., 2005, pp. 58–67.

18. Prasanth, K. V.; Sacco-Bubulya, P. A.; Prasanth, S. G.; Spector, D. L. (2003). Sequential entry of components of the gene expression machinery into daughter nuclei. Molecular Biology of the Cell. **14** *(3): 1043–57. doi:10.1091/mbc.E02-10-0669. PMC 151578. PMID 12631722.*

19. Kadauke, S; Blobel, G. A. (2013). Mitotic bookmarking by transcription factors. Epigenetics & Chromatin. **6** *(1): 6. doi:10.1186/1756-8935-6-6. PMC 3621617. PMID 23547918.*

20. Prescott, D. M.; Bender, M. A. (1962). Synthesis of RNA and protein during mitosis in mammalian tissue culture cells. Experimental Cell Research. **26** *(2): 260–8. doi:10.1016/0014-4827(62)90176-3. PMID 14488623.*

21. Olson, Mark O. J. (2011). The Nucleolus. Volume 15 of Protein Reviews. Berlin: Springer Science & Business Media. p. 15. ISBN 9781461405146.

22. Basto, R.; Lau, J.; Vinogradova, T.; Gardiol, A.; Woods, G.; Khodjakov, A.; Raff, W. (June 2006). Flies without centrioles. Cell. **125** *(7): 1375–1386. doi:10.1016/j.cell.2006.05.025. PMID 16814722.*

23. Heywood, P. (1978). Ultrastructure of mitosis in the chloromonadophycean alga Vacuolaria virescens. Journal of Cell Science. **31**: 37–51. PMID 670329.

24. Ribeiro, K.; Pereira-Neves, A; Benchimol, M. (2002). The mitotic spindle and associated membranes in the closed mitosis of trichomonads. Biology of the Cell. **94** *(3): 157–72. doi:10.1016/S0248-4900(02)01191-7. PMID 12206655.*

25. Chan, G.; Liu, S.; Yen, T. (2005). Kinetochore structure and function. Trends in Cell Biology. **15** *(11): 589–98. doi:10.1016/j.tcb.2005.09.010. PMID 16214339.*

26. Cheeseman, Iain M.; Desai, Arshad (January 2008). Molecular architecture of the kinetochore–microtubule interface. Nature Reviews Molecular Cell Biology. **9** *(1): 33–46. doi:10.1038/nrm2310. PMID 18097444.*

27. Winey, M.; Mamay, C.; O›Toole, E.; Mastronarde, D.; Giddings, T.; McDonald, K.; McIntosh, J. (1995). Three-dimensional ultrastructural analysis of the Saccharomyces cerevisiae mitotic spindle. Journal of Cell Biology. **129** *(6): 1601–15. doi:10.1083/jcb.129.6.1601. PMC 2291174. PMID 7790357.*

28. Maiato, H.; DeLuca, J.; Salmon, E.; Earnshaw, W. (2004). The dynamic kinetochore-microtubule interface. Journal of Cell Science. **117** *(Pt 23): 5461–77. doi:10.1242/jcs.01536. PMID 15509863.*

29. Chan, G.; Yen, T. (2003). The mitotic checkpoint: a signaling pathway that allows a single unattached kinetochore to inhibit mitotic exit. Progress in Cell Cycle Research. **5**: 431–9. PMID 14593737.

30. FitzHarris, G. (March 2012). "Anaphase B precedes anaphase A in the mouse egg". Current Biology. **22** *(5): 437–444. doi:10.1016/j.cub.2012.01.041. PMID 22342753.*

31. Miller, K. R. g (2000). Anaphase. Biology (5 ed.). Pearson Prentice Hall. pp. 169–70. ISBN 978-0-13-436265-6.

32. Chromosome condensation through mitosis. Science Daily. Retrieved 12 June 2007.

33. Glotzer, M. (2005). The molecular requirements for cytokinesis. Science. **307** *(5716): 1735–9. doi:10.1126/science.1096896. PMID 15774750.*

34. Albertson, R.; Riggs, B.; Sullivan, W. (2005). Membrane traffic: a driving force in cytokinesis. Trends in Cell Biology. **15** *(2): 92–101. doi:10.1016/j.tcb.2004.12.008. PMID 15695096.*

35. Raven et al., 2005, pp. 64–7, 328–9.

36. Lilly, .M; Duronio, R. (2005). New insights into cell cycle control from the Drosophila endocycle. Oncogene. **24** *(17): 2765–75. doi:10.1038/sj.onc.1208610. PMID 15838513.*

37. Sauer, F. C. (1935). Mitosis in the neural tube. Journal of Comparative Neurology. **62** *(2): 377–405. doi:10.1002/cne.900620207.*

38. Meyer, E. J. (2011). Interkinetic Nuclear Migration Is a Broadly Conserved Feature of Cell Division in Pseudostratified Epithelia. Current Biology. **21** *(6): 485–491. doi:10.1016/j.cub.2011.02.002. PMID 21376598.*

39. Luxenburg, Chen; et al. (2011). Developmental roles for Srf, cortical cytoskeleton and cell shape in epidermal spindle orientation. Nature Cell Biology. **13** *(3): 203–U51. doi:10.1038/Ncb2163. PMID 21336301.*

40. Nakajima, Yu-Ichiro; Meyer, Emily J; Kroesen, Amanda; McKinney, Sean A; Gibson, Matthew C (2013). Epithelial junctions maintain tissue architecture by directing planar spindle orientation. Nature. **500** *(7462): 359–362. doi:10.1038/nature12335. PMID 23873041.*

41. Cadart, C.; Zlotek-Zlotkiewicz, E.; Le Berre, M.; Piel, M.; Matthews, H. K. (2014). Exploring the Function of Cell Shape and Size during Mitosis. Developmental Cell. **29** *(2): 159–169. doi:10.1016/j.devcel.2014.04.009. PMID 24780736.*

42. Lancaster, Oscar; et al. (2013). Mitotic rounding alters cell geometry to ensure efficient bipolar spindle formation. Developmental Cell. **25** *(2): 270–283. doi:10.1016/j.devcel.2013.03.014. PMID 23623611.*

43. Lancaster, Oscar; Baum, Buzz (2014). Shaping up to divide: Coordinating actin and microtubule cytoskeletal remodelling during mitosis. Seminars in Cell & Developmental Biology. **34**: 109–115. doi:10.1016/j.semcdb.2014.02.015. PMID 24607328.

44. Maddox, A. S.; Burridge, K. (2003). "RhoA is required for cortical retraction and rigidity during mitotic cell rounding". Journal of Cell Biology. **160** *(2): 255–265. doi:10.1083/jcb.200207130. PMID 12538643.*

45. Matthews, H. K.; et al. (2012). Changes in ect2 localization couple actomyosin-dependent cell shape changes to mitotic progression. Developmental Cell. **23** *(2): 371–383. doi:10.1016/j.devcel.2012.06.003. PMID 22898780.*
46. Stewart, M. P.; Helenius, J.; Toyoda, Y.; Ramanathan, S. P.; Muller, D. J.; Hyman, A. A. (2011). Hydrostatic pressure and the actomyosin cortex drive mitotic cell rounding. Nature. **469** *(7329): 226–230. doi:10.1038/nature09642. PMID 21196934.*
47. Fischer-Friedrich, Elizabeth; et al. (2014). Quantification of surface tension and internal pressure generated by single mitotic cells. Scientific Reports. **4** *(6213). doi:10.1038/srep06213. PMID 25169063.*
48. Ramanathan, S. P.; et al. (2015). Cdk1-dependent mitotic enrichment of cortical myosin II promotes cell rounding against confinement. Nature Cell Biology. **17** *(2): 148–159. doi:10.1038/ncb3098. PMID 25621953.*
49. Mantikou, E.; Wong, K.M.; Repping, S.; Mastenbroek, S. (December 2012). Molecular origin of mitotic aneuploidies in preimplantation embryos. Biochimica et Biophysica Acta (BBA) - Molecular Basis of Disease. **1822** *(12): 1921–1930. doi:10.1016/j.bbadis.2012.06.013. PMID 22771499.*
50. Draviam, V.; Xie, S.; Sorger, P. (2004). Chromosome segregation and genomic stability. Current Opinion in Genetics & Development. **14** *(2): 120–5. doi:10.1016/j.gde.2004.02.007. PMID 15196457.*
51. Iourov, I.Y.; Vorsanova, S.G.; Yurov, Y.B. (2006). Chromosomal Variations in Mammalian Neuronal Cells: Known Facts and Attractive Hypotheses. In Jeon, Kwang J. International Review Of Cytology: A Survey of Cell Biology. **249**. Waltham, MA: Academic Press. p. 146. ISBN 9780080463506.
52. Shi, Q.; King, R.W. (October 2005). Chromosome nondisjunction yields tetraploid rather than aneuploid cells in human cell lines. Nature. **437** *(7061): 1038–1042. doi:10.1038/nature03958. PMID 16222248.*
53. Edgar, B.A.; Orr-Weaver, T.L. (May 2001). Endoreplication Cell Cycles: More for Less. Cell. **105** *(3): 297–306. doi:10.1016/S0092-8674(01)00334-8. PMID 11348589.*
54. Lee, H.O.; Davidson, J.M.; Duronio, R.J. (2009). Endoreplication: polyploidy with purpose. Genes & Development. **23** *(21): 2461–2477. doi:10.1101/gad.1829209. PMC 2779750. PMID 19884253.*
55. Italiano, J. E.; Shivdasani, R. A. (2003). Megakaryocytes and beyond: the birth of platelets. Journal of Thrombosis and Haemostasis. **1** *(6): 1174–82. doi:10.1046/j.1538-7836.2003.00290.x. PMID 12871316.*
56. Vitrat, N.; Cohen-Solal, K.; Pique, C.; LeCouedic, J.P.; Norol, F.; Larsen, A.K.; Katz, A.; Vainchenker, W.; Debili, N. (May 1998). Endomitosis of Human Megakaryocytes Are Due to Abortive Mitosis. Blood. **91** *(10): 3711–3723. PMID 9573008.*
57. Conlon, Ian; Raff, Martin (1999). Size Control in Animal Development. Cell. 96 (2): 235–44. doi:10.1016/S0092-8674(00)80563-2. PMID 9988218.

58. Nigg, Erich A.; Meraldi, Patrick; Lukas, Jiri; Fry, Andrew M.; Bartek, Jiri (1999). "Centrosome duplication in mammalian somatic cells requires E2F and Cdk2-cyclin A". Nature Cell Biology. 1 (2): 88–93. doi:10.1038/10054. PMID 10559879.

59. Mayor, T; Meraldi, P; Stierhof, YD; Nigg, EA; Fry, AM (1999). "Protein kinases in control of the centrosome cycle". FEBS Letters. 452 (1–2): 92–5. doi:10.1016/S0014-5793(99)00534-7. PMID 10376685.

60. Morgan, David O. (2006-09-06). The Cell Cycle: Principles of Control (Primers in Biology) (1 ed.). New Science Press, Ltd. ISBN 0-87893-508-8.

61. Nicklas, R. B. (1997). "How Cells Get the Right Chromosomes". Science. 275 (5300): 632–7. doi:10.1126/science.275.5300.632. PMID 9005842.

62. Lončarek, Jadranka; Kisurina-Evgenieva, Olga; Vinogradova, Tatiana; Hergert, Polla; La Terra, Sabrina; Kapoor, Tarun M.; Khodjakov, Alexey (2007). "The centromere geometry essential for keeping mitosis error free is controlled by spindle forces". Nature. 450 (7170): 745–9. doi:10.1038/nature06344. PMC 2586812 . PMID 18046416.

63. Dewar, Hilary; Tanaka, Kozo; Nasmyth, Kim; Tanaka, Tomoyuki U. (2004). "Tension between two kinetochores suffices for their bi-orientation on the mitotic spindle". Nature. 428 (6978): 93–7. doi:10.1038/nature02328. PMID 14961024.

64. Cimini, Daniela; Wan, Xiaohu; Hirel, Christophe B.; Salmon, E.D. (2006). "Aurora Kinase Promotes Turnover of Kinetochore Microtubules to Reduce Chromosome Segregation Errors". Current Biology. 16 (17): 1711–8. doi:10.1016/j.cub.2006.07.022. PMID 16950108.

65. Gautschi, O.; Heighway, J.; Mack, P. C.; Purnell, P. R.; Lara, P. N.; Gandara, D. R. (2008). "Aurora Kinases as Anticancer Drug Targets". Clinical Cancer Research. 14 (6): 1639–48. doi:10.1158/1078-0432.CCR-07-2179. PMID 18347165.

66. Zirkle, R. E. (1970). "Ultraviolet-Microbeam Irradiation of Newt-Cell Cytoplasm: Spindle Destruction, False Anaphase, and Delay of True Anaphase". Radiation Research. 41 (3): 516–37. doi:10.2307/3572841. JSTOR 3572841. PMID 5438206.

67. Rieder, CL; Palazzo, RE (1992). "Colcemid and the mitotic cycle". Journal of Cell Science. 102 (3): 387–92. PMID 1506421.

68. Burke, Daniel J.; Stukenberg, P. Todd (2008). "Linking Kinetochore-Microtubule Binding to the Spindle Checkpoint". Developmental Cell. 14 (4): 474–9. doi:10.1016/j.devcel.2008.03.015. PMC 2696048 . PMID 18410725.

69. Li, Rong; Murray, Andrew W. (1991). "Feedback control of mitosis in budding yeast". Cell. 66 (3): 519–31. doi:10.1016/0092-8674(81)90015-5. PMID 1651172.

70. Hoyt, M.Andrew; Totis, Laura; Roberts, B.Tibor (1991). "S. cerevisiae genes required for cell cycle arrest in response to loss of microtubule function". Cell. 66 (3): 507–17. doi:10.1016/0092-8674(81)90014-3. PMID 1651171.

71. Neff, MW; Burke, DJ (1992). "A delay in the Saccharomyces cerevisiae cell cycle that is induced by a dicentric chromosome and dependent upon mitotic checkpoints". Molecular and Cellular Biology. 12 (9): 3857–64. PMC 360258

. PMID 1324407.

72. Wells, W. A.; Murray, AW (1996). "Aberrantly segregating centromeres activate the spindle assembly checkpoint in budding yeast". The Journal of Cell Biology. 133 (1): 75–84. doi:10.1083/jcb.133.1.75. PMC 2120768 . PMID 8601615.
73. Hardwick, K. G.; Weiss, E.; Luca, F. C.; Winey, M.; Murray, A. W. (1996). "Activation of the Budding Yeast Spindle Assembly Checkpoint Without Mitotic Spindle Disruption". Science. 273 (5277): 953–6. doi:10.1126/science.273.5277.953. PMID 8688079.
74. Wang, Y; Burke, DJ (1995). "Checkpoint genes required to delay cell division in response to nocodazole respond to impaired kinetochore function in the yeast Saccharomyces cerevisiae". Molecular and Cellular Biology. 15 (12): 6838–44. PMC 230938 . PMID 8524250.
75. Spencer, F.; Hieter, P. (1992). "Centromere DNA Mutations Induce a Mitotic Delay in Saccharomyces cerevisiae". Proceedings of the National Academy of Sciences of the United States of America. 89 (19): 8908–8912. doi:10.1073/pnas.89.19.8908. JSTOR 2360300. PMC 50033 . PMID 1409584.
76. Hardwick, KG; Li, R; Mistrot, C; Chen, RH; Dann, P; Rudner, A; Murray, AW (1999). "Lesions in many different spindle components activate the spindle checkpoint in the budding yeast Saccharomyces cerevisiae". Genetics. 152 (2): 509–18. PMC 1460633 . PMID 10353895.
77. McIntosh, JR (1991). "Structural and mechanical control of mitotic progression". Cold Spring Harbor symposia on quantitative biology. 56: 613–9. doi:10.1101/sqb.1991.056.01.070. PMID 1819511.
78. Rieder, C. L.; Cole, RW; Khodjakov, A; Sluder, G (1995). "The checkpoint delaying anaphase in response to chromosome monoorientation is mediated by an inhibitory signal produced by unattached kinetochores". The Journal of Cell Biology. 130 (4): 941–8. doi:10.1083/jcb.130.4.941. PMC 2199954 . PMID 7642709.
79. Li, X; Nicklas, RB (1997). "Tension-sensitive kinetochore phosphorylation and the chromosome distribution checkpoint in praying mantid spermatocytes". Journal of Cell Science. 110 (5): 537–45. PMID 9092936.
80. Rieder, C. L.; Khodjakov, A.; Paliulis, L. V.; Fortier, T. M.; Cole, R. W.; Sluder, G. (1997). "Mitosis in vertebrate somatic cells with two spindles: Implications for the metaphase/anaphase transition checkpoint and cleavage". Proceedings of the National Academy of Sciences. 94 (10): 5107–12. doi:10.1073/pnas.94.10.5107.
81. Hirano, Tatsuya (2000). "CHROMOSOMECOHESION, CONDENSATION, ANDSEPARATION". Annual Review of Biochemistry. 69: 115–44. doi:10.1146/annurev.biochem.69.1.115. PMID 10966455.
82. Tanaka, K.; Hao, Z; Kai, M; Okayama, H (2001). "Establishment and maintenance of sister chromatid cohesion in fission yeast by a unique mechanism". The EMBO Journal. 20 (20): 5779–90. doi:10.1093/emboj/20.20.5779. PMC 125673 . PMID 11598020.

83. Gonzalez, Cayetano; Jimenez, Jose Casal; Ripoll, Pedro; Sunkel, Claudio E. (1991). "The spindle is required for the process of sister chromatid separation in Drosophila neuroblasts". Experimental Cell Research. 192 (1): 10–5. doi:10.1016/0014-4827(91)90150-S. PMID 1898588.

84. Losada, Ana; Hirano, Tatsuya (2001). "Shaping the metaphase chromosome: coordination of cohesion and condensation". BioEssays. 23 (10): 924–35. doi:10.1002/bies.1133. PMID 11598959.

85. Bernard, P. (2001). "Requirement of Heterochromatin for Cohesion at Centromeres". Science. 294 (5551): 2539–542. doi:10.1126/science.1064027.

86. Nonaka, Nobuhiro; Kitajima, Tomoya; Yokobayashi, Shihori; Xiao, Guoping; Yamamoto, Masayuki; Grewal, Shiv I. S.; Watanabe, Yoshinori (2001). "Recruitment of cohesin to heterochromatic regions by Swi6/HP1 in fission yeast". Nature Cell Biology. 4 (1): 89–93. doi:10.1038/ncb739. PMID 11780129.

87. Hall, I. M.; Noma, K.-i.; Grewal, S. I. S. (2002). "RNA interference machinery regulates chromosome dynamics during mitosis and meiosis in fission yeast". Proceedings of the National Academy of Sciences. 100: 193–8. doi:10.1073/pnas.232688099.

88. Fukagawa, Tatsuo; Nogami, Masahiro; Yoshikawa, Mitsuko; Ikeno, Masashi; Okazaki, Tuneko; Takami, Yasunari; Nakayama, Tatsuo; Oshimura, Mitsuo (2004). "Dicer is essential for formation of the heterochromatin structure in vertebrate cells". Nature Cell Biology. 6 (8): 784–91. doi:10.1038/ncb1155. PMID 15247924.

89. Weber, Stewart A.; Gerton, Jennifer L.; Polancic, Joan E.; Derisi, Joseph L.; Koshland, Douglas; Megee, Paul C. (2004). "The Kinetochore Is an Enhancer of Pericentric Cohesin Binding". PLoS Biology. 2 (9): e260. doi:10.1371/journal.pbio.0020260. PMC 490027 . PMID 15309047.

90. Prasanth, Supriya G; Prasanth, Kannanganattu V; Siddiqui, Khalid; Spector, David L; Stillman, Bruce (2004). "Human Orc2 localizes to centrosomes, centromeres and heterochromatin during chromosome inheritance". The EMBO Journal. 23 (13): 2651–63. doi:10.1038/sj.emboj.7600255. PMC 449767 . PMID 15215892.

91. Shimada, Kenji; Gasser, Susan M. (2007). "The Origin Recognition Complex Functions in Sister-Chromatid Cohesion in Saccharomyces cerevisiae". Cell. 128 (1): 85–99. doi:10.1016/j.cell.2006.11.045. PMID 17218257.

92. Kato, H; Matsunaga, F; Miyazaki, S; Yin, L; D'urso, G; Tanaka, K; Murakami, Y (2008). "Schizosaccharomyces pombe Orc5 plays multiple roles in the maintenance of genome stability throughout the cell cycle". Cell cycle. 7 (8): 1085–96. doi:10.4161/cc.7.8.5710. PMID 18414064.

93. Hauf, Silke; Watanabe, Yoshinori (2004). "Kinetochore Orientation in Mitosis and Meiosis". Cell. 119 (3): 317–27. doi:10.1016/j.cell.2004.10.014. PMID 15507205.

94. Sonoda, Eiichiro; Matsusaka, Takahiro; Morrison, Ciaran; Vagnarelli, Paola; Hoshi, Osamu; Ushiki, Tatsuo; Nojima, Kuniharu; Fukagawa, Tatsuo; et al. (2001). "Scc1/Rad21/Mcd1 Is Required for Sister Chromatid Cohesion and

Kinetochore Function in Vertebrate Cells". Developmental Cell. 1 (6): 759–70. doi:10.1016/S1534-5807(01)00088-0. PMID 11740938.

95. Vass, Sharron; Cotterill, Sue; Valdeolmillos, Ana M.; Barbero, José L.; Lin, Enmoore; Warren, William D.; Heck, Margarete M.S. (2003). "Depletion of Drad21/Scc1 in Drosophila Cells Leads to Instability of the Cohesin Complex and Disruption of Mitotic Progression". Current Biology. 13 (3): 208–18. doi:10.1016/S0960-9822(03)00047-2. PMID 12573216.
96. Haering, Christian H.; Löwe, Jan; Hochwagen, Andreas; Nasmyth, Kim (2002). "Molecular Architecture of SMC Proteins and the Yeast Cohesin Complex". Molecular Cell. 9 (4): 773–88. doi:10.1016/S1097-2765(02)00515-4. PMID 11983169.
97. Hirano, T. (1999). "SMC-mediated chromosome mechanics: a conserved scheme from bacteria to vertebrates?". Genes & Development. 13: 11–9. doi:10.1101/gad.13.1.11.
98. Ciosk, Rafal; Zachariae, Wolfgang; Michaelis, Christine; Shevchenko, Andrej; Mann, Matthias; Nasmyth, Kim (1998). "An ESP1/PDS1 Complex Regulates Loss of Sister Chromatid Cohesion at the Metaphase to Anaphase Transition in Yeast". Cell. 93 (6): 1067–76. doi:10.1016/S0092-8674(00)81211-8. PMID 9635435.
99. Alexandru, Gabriela; Uhlmann, Frank; Mechtler, Karl; Poupart, Marc-André; Nasmyth, Kim (2001). "Phosphorylation of the Cohesin Subunit Scc1 by Polo/Cdc5 Kinase Regulates Sister Chromatid Separation in Yeast". Cell. 105 (4): 459–72. doi:10.1016/S0092-8674(01)00362-2. PMID 11371343.
100. Leismann, O. (2000). "Degradation of Drosophila PIM regulates sister chromatid separation during mitosis". Genes & Development. 14 (17): 2192–205. doi:10.1101/gad.176700.
101. Zur, A.; Brandeis, M (2001). "Securin degradation is mediated by fzy and fzr, and is required for complete chromatid separation but not for cytokinesis". The EMBO Journal. 20 (4): 792–801. doi:10.1093/emboj/20.4.792. PMC 145417 . PMID 11179223.
102. Sumara, I.; Vorlaufer, E.; Gieffers, C.; Peters, B. H.; Peters, J.-M. (2000). "Characterization of Vertebrate Cohesin Complexes and Their Regulation in Prophase". The Journal of Cell Biology. 151 (4): 749–62. doi:10.1083/jcb.151.4.749. PMC 2169443 . PMID 11076961.
103. Losada, A.; Yokochi, T.; Kobayashi, R.; Hirano, T. (2000). "Identification and Characterization of Sa/Scc3p Subunits in the Xenopus and Human Cohesin Complexes". The Journal of Cell Biology. 150 (3): 405–16. doi:10.1083/jcb.150.3.405. PMC 2175199 . PMID 10931856.
104. Giménez-Abián, Juan F; Sumara, Izabela; Hirota, Toru; Hauf, Silke; Gerlich, Daniel; De La Torre, Consuelo; Ellenberg, Jan; Peters, Jan-Michael (2004). "Regulation of Sister Chromatid Cohesion between Chromosome Arms". Current Biology. 14 (13): 1187–93. doi:10.1016/j.cub.2004.06.052. PMID 15242616.
105. Paliulis, Leocadia V.; Nicklas, R.Bruce (2004). "Micromanipulation of

Chromosomes Reveals that Cohesion Release during Cell Division Is Gradual and Does Not Require Tension". Current Biology. 14 (23): 2124–9. doi:10.1016/j.cub.2004.11.052. PMID 15589155.

106. Nakajima, M.; Kumada, K.; Hatakeyama, K.; Noda, T.; Peters, J.-M.; Hirota, T. (2007). "The complete removal of cohesin from chromosome arms depends on separase". Journal of Cell Science. 120 (23): 4188. doi:10.1242/jcs.011528.

107. McGuinness, Barry E.; Hirota, Toru; Kudo, Nobuaki R.; Peters, Jan-Michael; Nasmyth, Kim (2005). "Shugoshin Prevents Dissociation of Cohesin from Centromeres During Mitosis in Vertebrate Cells". PLoS Biology. 3 (3): e86. doi:10.1371/journal.pbio.0030086. PMC 1054882 . PMID 15737064.

108. Salic, Adrian; Waters, Jennifer C.; Mitchison, Timothy J. (2004). "Vertebrate Shugoshin Links Sister Centromere Cohesion and Kinetochore Microtubule Stability in Mitosis". Cell. 118 (5): 567–78. doi:10.1016/j.cell.2004.08.016. PMID 15339662.

109. De Antoni, Anna; Pearson, Chad G.; Cimini, Daniela; Canman, Julie C.; Sala, Valeria; Nezi, Luigi; Mapelli, Marina; Sironi, Lucia; et al. (2005). "The Mad1/Mad2 Complex as a Template for Mad2 Activation in the Spindle Assembly Checkpoint". Current Biology. 15 (3): 214–25. doi:10.1016/j.cub.2005.01.038. PMID 15694304.

110. Musacchio, Andrea; Edward D. Salmon (May 2007). "The spindle-assembly checkpoint in space and time". Nat Rev Mol Cell Biol. 8 (5): 379–393. doi:10.1038/nrm2163. ISSN 1471-0072. PMID 17426725.

111. Martin-Lluesma, S.; Stucke, VM; Nigg, EA (2002). "Role of Hec1 in Spindle Checkpoint Signaling and Kinetochore Recruitment of Mad1/Mad2". Science. 297 (5590): 2267–70. doi:10.1126/science.1075596. PMID 12351790.

112. Lens, Susanne M.A.; Rob M.F. Wolthuis; Rob Klompmaker; Jos Kauw; Reuven Agami; Thijn Brummelkamp; Geert Kops; René H. Medema (2003-06-16). "Survivin is required for a sustained spindle checkpoint arrest in response to lack of tension". The EMBO Journal. 22 (12): 2934–2947. doi:10.1093/emboj/cdg307. ISSN 0261-4189. PMC 162159 . PMID 12805209.

113. Hauf, S; Cole, R. W.; Laterra, S; Zimmer, C; Schnapp, G; Walter, R; Heckel, A; Van Meel, J; Rieder, C. L.; Peters, J. M. (2003). "The small molecule Hesperadin reveals a role for Aurora B in correcting kinetochore–microtubule attachment and in maintaining the spindle assembly checkpoint". The Journal of Cell Biology. 161 (2): 281–294. doi:10.1083/jcb.200208092. PMC 2172906 . PMID 12707311.

114. Aravamudhan, Pavithra; Goldfarb, Alan A.; Joglekar, Ajit P. "The kinetochore encodes a mechanical switch to disrupt spindle assembly checkpoint signalling". Nature Cell Biology. 17 (7): 868–879. doi:10.1038/ncb3179. PMC 4630029 . PMID 26053220.

115. Alberts, B., Johnson, A., Lewis, J., Morgan, D., Raff, M., Roberts, K., & Walter, P. (2015). Molecular Biology of The Cell (6th ed.). New York, NY: Garland Science, Taylor & Francis Group. p. 988. ISBN 978-0-8153-4432-2.

116. Kops, Geert J.P.L.; Weaver, Beth A. A.; Cleveland, Don W. (2005). "On the road to cancer: aneuploidy and the mitotic checkpoint". Nature Reviews Cancer. 5 (10): 773–785. doi:10.1038/nrc1714. PMID 16195750.

117. Lengauer, Christoph; Kinzler, Kenneth W.; Vogelstein, Bert (1998). "Genetic instabilities in human cancers". Nature. 396 (6712): 643–649. doi:10.1038/25292. PMID 9872311.

118. Weaver, Beth AA; Cleveland, Don W (2006). "Does aneuploidy cause cancer?". Current Opinion in Cell Biology. 18 (6): 658–67. doi:10.1016/j.ceb.2006.10.002. PMID 17046232.

119. Lengauer, Christoph; Cahill, Daniel P.; Yu, Jian; Riggins, Gregory J.; Willson, James K. V.; Markowitz, Sanford D.; Kinzler, Kenneth W.; Vogelstein, Bert (1998). "Mutations of mitotic checkpoint genes in human cancers". Nature. 392 (6673): 300–3. doi:10.1038/32688. PMID 9521327.

120. Diaz-Rodriguez, Elena; Álvarez-Fernández, Stela; Chen, Xi; Paiva, Bruno; López-Pérez, Ricardo; García-Hernández, Juan Luis; San Miguel, Jesús F.; Pandiella, Atanasio (2011). "Deficient Spindle Assembly Checkpoint in Multiple Myeloma". PLoS ONE. 6 (11): e27583. doi:10.1371/journal.pone.0027583. PMC 3223182 . PMID 22132115.

121. Grady, William M. (2004). "Genomic instability and colon cancer". Cancer and Metastasis Review. 23 (1–2): 11–27. doi:10.1023/A:1025861527711.

122. Wang, Rui-Hong; Hongtao, Yu; Deng, Chu-Xia (2004). "A requirement for breast-cancer associated gene (BRCA1) in the spindle checkpoint". PNAS. 101 (49): 17108–17113. doi:10.1073/pnas.0407585101. PMC 535394 . PMID 15563594.

123. Sotillo, Rocío; Hernando, Eva; Díaz-Rodríguez, Elena; Teruya-Feldstein, Julie; Cordón-Cardo, Carlos; Lowe, Scott W.; Benezra, Robert (2007). "Mad2 Overexpression Promotes Aneuploidy and Tumorigenesis in Mice". Cancer Cell. 11 (1): 9–23. doi:10.1016/j.ccr.2006.10.019. PMC 1850996 . PMID 17189715.

124. Yamamoto, Y; Matsuyama, H; Chochi, Y; Okuda, M; Kawauchi, S; Inoue, R; Furuya, T; Oga, A; et al. (2007). "Overexpression of BUBR1 is associated with chromosomal instability in bladder cancer". Cancer Genetics and Cytogenetics. 174 (1): 42–7. doi:10.1016/j.cancergencyto.2006.11.012. PMID 17350465.

125. Weaver, B. A.A.; Cleveland, D. W. (2009). "The role of aneuploidy in promoting and suppressing tumors". The Journal of Cell Biology. 185 (6): 935–7. doi:10.1083/jcb.200905098. PMC 2711620 . PMID 19528293.

126. Cross, Shawn M.; Sanchez, Carissa A; Morgan, Catherine A.; Schimke, Melana K.; Reid, Brian J. (1995). "A p53-dependant mouse spindle checkpoint". Science. 3 (5202): 1353–1356. doi:10.1126/science.7871434.

127. Altieri, Dario C. (2001). "The molecular basis and potential role of survivin in cancer diagnosis and therapy". Trends in Molecular Medicine. 7 (12): 542–547. doi:10.1016/S1471-4914(01)02243-2. PMID 11733216.

128. Gisselsson, David; Hakanson, Ulf; Stoller, Patrick; Marti, Dominik; Jin, Yuesheng; Rosengren, Anders H.; Stewénius, Ylva; Kahl, Fredrik; Panagopoulos, Ioannis (2008). "When the Genome Plays Dice: Circumvention of the Spindle Assembly

Checkpoint and Near-Random Chromosome Segregation in Multipolar Cancer Cell Mitoses". PLoS ONE. 3 (4): e1871. doi:10.1371/journal.pone.0001871. PMC 2289843 . PMID 18392149.

129. Zhou, Jun; Giannakakou, Paraskevi (2005). "Targeting Microtubules for Cancer Chemotherapy". Current Medicinal Chemistry - Anti-Cancer Agents. 5 (1): 65–71. doi:10.2174/1568011053352569. PMID 15720262.
130. Carvajal, Richard D.; Tse, Archie; Schwartz, Gary K. (2006). "Aurora Kinases: New Targets for Cancer Therapy". Clinical Cancer Research. 12 (23): 6869–75. doi:10.1158/1078-0432.CCR-06-1405. PMID 17145803.
131. Anand, Shubha; Penrhyn-Lowe, Sue; Venkitaraman, Ashok R. (2003). "Aurora-A amplification overrides the mitotic spindle assembly checkpoint, inducing resistance to Taxol". Cancer Cell. 3 (1): 51–62. doi:10.1016/S1535-6108(02)00235-0. PMID 12559175.
132. Harrington, E. A.; Bebbington, D; Moore, J; Rasmussen, R. K.; Ajose-Adeogun, A. O.; Nakayama, T; Graham, J. A.; Demur, C; Hercend, T; Diu-Hercend, A; Su, M; Golec, J. M.; Miller, K. M.; et al. (2004). "VX-680, a potent and selective small-molecule inhibitor of the Aurora kinases, suppresses tumor growth in vivo". Nature Medicine. 10 (3): 262–267. doi:10.1038/nm1003. PMID 14981513.
133. Altieri, Dario C. (2008). "Survivin, cancer networks and pathway-directed drug discovery". Nature Reviews Cancer. 8 (1): 61–70. doi:10.1038/nrc2293. PMID 18075512.
134. Tao, Weikang; et al. (2005). "Induction of apoptosis by an inhibitor of the mitotic kinesin KSP requires both activation of the spindle assembly checkpoint and mitotic slippage". Cancer Cell. 8 (1): 49–59. doi:10.1016/j.ccr.2005.06.003. PMID 16023598.
135. Julian Lewis ... [et (2007). Molecular biology of the cell (5th ed.). New York: Garland Science. ISBN 9780815341055.
136. Cooper, Geoffrey M. (2000). The cell : a molecular approach (2nd ed.). Washington (DC): ASM Press. ISBN 0-87893-106-6.
137. al.], Harvey Lodish ... [et (2000). Molecular cell biology (4th ed.). New York: Scientific American Books. ISBN 0-7167-3136-3.
138. Malumbres, Marcos; Barbacid, Mariano (March 2009). "Cell cycle, CDKs and cancer: a changing paradigm". Nature Reviews Cancer. 9 (3): 153–166. doi:10.1038/nrc2602. PMID 19238148.
139. Vermeulen, Katrien; Van Bockstaele, Dirk R.; Berneman, Zwi N. (June 2003). "The cell cycle: a review of regulation, deregulation and therapeutic targets in cancer". Cell Proliferation. 36 (3): 131–149. doi:10.1046/j.1365-2184.2003.00266.x.
140. Bertoli, Cosetta; Skotheim, Jan M.; de Bruin, Robertus A. M. (23 July 2013). "Control of cell cycle transcription during G1 and S phases". Nature Reviews Molecular Cell Biology. 14 (8): 518–528. doi:10.1038/nrm3629. PMID 23877564.
141. Skotheim, Jan M.; Di Talia, Stefano; Siggia, Eric D.; Cross, Frederick R. (17 July 2008). "Positive feedback of G1 cyclins ensures coherent cell cycle entry".

Nature. 454 (7202): 291–296. doi:10.1038/nature07118. PMC 2606905Freely accessible. PMID 18633409.

142. Bartek, Jiri; Lukas, Jiri (December 2001). "Mammalian G1- and S-phase checkpoints in response to DNA damage". Current Opinion in Cell Biology. 13 (6): 738–747. doi:10.1016/S0955-0674(00)00280-5.
143. Bertoli, Cosetta; de Bruin, Robertus Antonius Maria (1 July 2014). "Turning cell cycle entry on its head". eLife. 3. doi:10.7554/eLife.03475.
144. Morgan, David (2007). The Cell Cycle Principles of Control. New Science Press. pp. 92–95.
145. Morgan, David (2007). The Cell Cycle Principles of Control. New Science Press. pp. 228–229.
146. Guardavaccaro, Daniele; Pagano, Michele (April 2006). "Stabilizers and Destabilizers Controlling Cell Cycle Oscillators". Molecular Cell. 22 (1): 1–4. doi:10.1016/j.molcel.2006.03.017.
147. Seki, A.; Coppinger, J. A.; Jang, C.-Y.; Yates, J. R.; Fang, G. (20 June 2008). "Bora and the Kinase Aurora A Cooperatively Activate the Kinase Plk1 and Control Mitotic Entry". Science. 320 (5883): 1655–1658. doi:10.1126/science.1157425.
148. Morgan, David (2007). The Cell Cycle Principles of Control. New Science Press. pp. 44–45, 90.
149. Sha, Wei; et al. (September 2002). "Hysteresis drives cell-cycle transitions in Xenopus laevis egg extracts". Proceedings of the National Academy of Sciences.
150. Wang, Yingmei; Ji, Ping; Liu, Jinsong; Broaddus, Russell R; Xue, Fengxia; Zhang, Wei (2009). "Centrosome-associated regulators of the G2/M checkpoint as targets for cancer therapy". Molecular Cancer. 8 (1): 8. doi:10.1186/1476-4598-8-8.
151. Löbrich, Markus; Jeggo, Penny A. (November 2007). "The impact of a negligent G2/M checkpoint on genomic instability and cancer induction". Nature Reviews Cancer. 7 (11): 861–869. doi:10.1038/nrc2248.
152. Harper, J. Wade; Elledge, Stephen J. (December 2007). "The DNA Damage Response: Ten Years After". Molecular Cell. 28 (5): 739–745. doi:10.1016/j.molcel.2007.11.015. PMID 18082599.
153. Peters, Jan-Michael (1998). "SCF and APC: the Yin and Yang of cell cycle regulated proteolysis". Current Opinion in Cell Biology. 10 (6): 759–68. doi:10.1016/S0955-0674(98)80119-1. PMID 9914180.
154. Ciosk, Rafal; Zachariae, Wolfgang; Michaelis, Christine; Shevchenko, Andrej; Mann, Matthias; Nasmyth, Kim (1998). "An ESP1/PDS1 Complex Regulates Loss of Sister Chromatid Cohesion at the Metaphase to Anaphase Transition in Yeast". Cell. 93 (6): 1067–76. doi:10.1016/S0092-8674(00)81211-8. PMID 9635435.
155. Karp, Gerald (2005). Cell and Molecular Biology: Concepts and Experiments (4th ed.). Hoboken, New Jersey: John Wiley and Sons. pp. 598–9. ISBN 0-471-16231-0.

156. Kastan, MB; Bartek, J (18 November 2004). "Cell-cycle checkpoints and cancer.". Nature. 432 (7015): 316–23. doi:10.1038/nature03097. PMID 15549093.

157. Shiloh, Y; Kastan, MB (2001). "ATM: genome stability, neuronal development, and cancer cross paths.". Advances in cancer research. 83: 209–54. doi:10.1016/s0065-230x(01)83007-4. PMID 11665719.

158. Lam, Michael H; Liu, Qinghua; Elledge, Stephen J; Rosen, Jeffrey M (July 2004). "Chk1 is haploinsufficient for multiple functions critical to tumor suppression". Cancer Cell. 6 (1): 45–59. doi:10.1016/j.ccr.2004.06.015.

159. King, MC; Marks, JH; Mandell, JB; New York Breast Cancer Study, Group (24 October 2003). "Breast and ovarian cancer risks due to inherited mutations in BRCA1 and BRCA2.". Science. 302 (5645): 643–6. doi:10.1126/science.1088759. PMID 14576434.

160. Venkitaraman, AR (25 January 2002). "Cancer susceptibility and the functions of BRCA1 and BRCA2.". Cell. 108 (2): 171–82. doi:10.1016/s0092-8674(02)00615-3. PMID 11832208.

CHAPTER 4

Maturation Promoting Factor

Chapter Contents

Maturation promoting factor (MPF) is a cell cycle checkpoint that regulates the passage of a cell from the G2 growth phase to the M phase. It is also known as the G2 checkpoint, and ensures that DNA replication during the S phase did not produce any mistakes. It also makes sure that the cell grew enough during the G2 stage. If the processes in the S and G2 phases went according to plan, and no abnormalities are present within the cell, then MPF will allow the cell to progress to the M phase, where it will make a copy of itself.

INTRODUCTION

Maturation-promoting factor (abbreviated **MPF**, also called **mitosis-promoting factor** or **M-Phase-promoting factor**) is the cyclin-Cdk complex that was discovered first in frog eggs. It stimulates the mitotic and meiotic phases of the cell cycle. MPF promotes the entrance into mitosis (the M phase) from the G_2 phase by phosphorylating multiple proteins needed during mitosis. MPF is activated at the end of G_2 by a phosphatase, which removes an inhibitory phosphate group added earlier.

The MPF is also called the M phase kinase because of its ability to phosphorylate target proteins at a specific point in the cell cycle and thus control their ability to function.

DISCOVERY

In 1971, two independent teams of researchers (Yoshio Masui and Clement Markert, as well as Dennis Smith and Robert Ecker) found that frog oocytes arrested in G_2 could be induced to enter M phase by microinjection of cytoplasm from oocytes that had been hormonally stimulated with progesterone. Because the entry of oocytes into meiosis is frequently referred to as oocyte maturation, this cytoplasmic factor was called maturation promoting factor (MPF). Further studies showed, however, that the activity of MPF is not restricted to the entry of oocytes into meiosis. To the contrary, MPF is also present in somatic cells, where it induces entry into M phase of the mitotic cycle.

Evidence that a diffusible factor regulates the entry into mitosis had been previously obtained in 1966 using the slime mold *Physarum polycephalum* in which the nuclei of the multi-nucleate plasmodial form undergo synchronous mitoses. Fusing plasmodia whose cell cycles were out of phase with each other led to a synchronous mitosis in the next mitotic cycle. This result demonstrated that mitotic entry was controlled by a diffusible cytoplasmic factor and not by a nuclear clock.

STRUCTURE

MPF is composed of two subunits:

- The cyclin-dependent kinase subunit, that uses ATP to phosphorylate specific serine and threonine residues of target proteins.
- Cyclin, a regulatory subunit. The cyclins are necessary for the kinase subunit to function with the appropriate substrate. The mitotic cyclins can be grouped as cyclins A & B. These cyclins have a nine residue sequence in the N-terminal region called the "destruction box", which can be recognized by the ubiquitin ligase enzyme which destroys the cyclins when appropriate.

Role in the cell cycle

During G_1 and S phase, the CDK1 subunit of MPF is inactive due to an inhibitory enzyme, Wee1. Wee1 phosphorylates the Thr-14 residues in yeast and Tyr-15 residues in humans of CDK1, rendering MPF inactive. During the transition of G_2 to M phase, cdk1 is de-phosphorylated by CDC25. The CDK1 subunit is now free and can bind to cyclin B, activate MPF, and make the cell enter mitosis. There is also a positive feedback loop that inactivates wee1.

Activation of MPF

MPF must be activated in order for the cell to transition from G_2 to M phase. There are three amino acid residues responsible for this G_2 to M phase transition. The Threonine-161

(Thr-161) on CDK1 must be phosphorylated by a Cyclin Activating Kinase (CAK). CAK only phosphorylates Thr-161 when cyclin B is attached to CDK1.

In addition, two other residues on the CDK1 subunit must be activated by dephosphorylation. CDC25 removes a phosphate from residues Threonine-14 (Thr-14) and Tyrosine-15 (Tyr-15) and adds a hydroxyl group. Cyclin B/CDK1 activates CDC25 resulting in a positive feedback loop.

Overview of Functions of MPF

- Triggers the formation of mitotic spindle through microtubule instability.
- Promotes mitosis i.e. chromatin condensation through phosphorylation of condensins.
- The three lamins present in the nuclear lamina, lamin A, B & C, are phosphorylated by MPF at serine amino residues. This leads to depolymerisation of the nuclear lamina & breakdown of nuclear envelope into small vesicles.
- Causes phosphorylation of GM130, which leads to the fragmentation of the Golgi and the ER

Targets of MPF

The following are affected by MPF.

- condensins, which enable chromatin condensation (see prophase)
- various microtubule-associated proteins involved in mitotic spindle formation
- lamins, interaction contributing to degradation of the nuclear envelope
- Histones, H_1 and H_3
- Golgi matrix, to cause fragmentation

Inhibition of Myosin

MPF phosphorylates inhibitory sites on myosin early in mitosis. This prevents cytokinesis. When MPF activity falls at anaphase, the inhibitory sites are dephosphorylated and cytokinesis proceeds.

Disassembly by Anaphase-Promoting Complex

MPF is disassembled when anaphase-promoting complex (APC) polyubiquitinates cyclin B, marking it for degradation in a negative feedback loop. In intact cells, cyclin degradation begins shortly after the onset of anaphase (late anaphase), the period of mitosis when sister chromatids are separated and pulled toward opposite spindle poles. As the concentration of Cyclin B/CDK1 increases, the heterodimer promotes APC to polyubiquitinate Cyclin B/CDK1.

REFERENCES

1. Smith, L.Dennis; Ecker, R.E. (May 1971). The interaction of steroids with Rana pipiens oocytes in the induction of maturation. Developmental Biology. 25 (2): 232–247. doi:10.1016/0012-1606(71)90029-7.
2. Masui, Yoshio; Markert, Clement L. (May 1971). Cytoplasmic control of nuclear behavior during meiotic maturation of frog oocytes. Journal of Experimental Zoology. 177 (2): 129–145. doi:10.1002/jez.1401770202.
3. Masui Y, Markert CL (1971), Cytoplasmic control of nuclear behavior during meiotic maturation of frog oocytes. J Exp Zool 177:2, 129-145
4. Smith LD, Ecker RE (1971), The interaction of steroids with Rana pipiens oocytes in the induction of maturation. Dev Biol 25:2, 232-247
5. Rusch HP, Sachsenmaier W, Behrens K, Gruter V. (1966). Synchronization of mitosis by the fusion of the plasmodia of Physarum polycephalum. J Cell Biol.31:204-9.

CHAPTER 5

Mitotic Bookmarking

Chapter Contents

The emerging perspective suggests that histone modifications play a role in gene bookmarking during mitosis and in inheritance of chromatin state during differentiation. Histone modifications often work in concert with other epigenetic marks (e.g. DNA methylation) to ensure the inheritance of specific gene expression patterns through cell generations.

INTRODUCTION

Bookmarking (also "**gene bookmarking**" or "**mitotic bookmarking**") refers to a potential mechanism of transmission of gene expression programs through cell division.

During mitosis, gene transcription is silenced and most transcription factors are removed from chromatin. The term bookmarking compares transcription to reading from a book. The pause in transcription during mitosis is like closing the book. Molecular bookmarks are the factors that allow transcription to resume in an orderly fashion in newborn cells following mitosis (when the book is re-opened).

Bookmarks fulfill the following criteria:

- at some point prior to the onset of mitosis, the promoters of genes that exist in a transcription-competent state become marked in some way,
- this mark persists both during and after mitosis, and
- the marking transmits gene expression memory by preventing the mitotic compaction of DNA at this locus, or by facilitating reassembly of transcription complexes on the promoter, or both.

BOOKMARKING AS SELECTIVE MAINTENANCE OF AN OPEN CHROMATIN STATE

The term bookmarking was originally coined to describe the non-compaction of some gene promoters during mitosis. More recently genome accessibility during interphase and mitosis has been directly compared on a genome-wide scale. While gene promoters tend to be better preserved than distal regulatory elements, substantial variation exists at individual sites. Bookmarking by DNA methylation, histone post-translation modification, and histone variants

Some patterns of histone modification and the presence of some histone variants on DNA remain unchanged during mitosis and have the potential to act as bookmarks. Patters of DNA methylation are generally unchanged and also have the potential to function as bookmarks.

BOOKMARKING BY TRANSCRIPTION FACTORS

Notable transcription factors implicated in mitotic bookmarking include:

- BRD4
- GATA1
- HSP70
- TFIID
- ESRRB

Hsp70

The **70 kilodalton heat shock proteins** (**Hsp70**s or **DnaK**) are a family of conserved ubiquitously expressed heat shock proteins. Proteins with similar structure exist in virtually all living organisms. The Hsp70s are an important part of the cell's machinery for protein folding, and help to protect cells from stress.

Discovery

Members of the Hsp70 family are very strongly upregulated by heat stress and toxic chemicals, particularly heavy metals such as arsenic, cadmium, copper, mercury, etc. Hsp70 was originally discovered by FM Ritossa in the 1960s when a lab worker accidentally boosted the incubation temperature of Drosophila (fruit flies). When examining the chromosomes, Ritossa found a "puffing pattern" that indicated the elevated gene transcription of an unknown protein. This was later described as the "Heat Shock Response" and the proteins were termed the "Heat Shock Proteins" (Hsps).

Structure

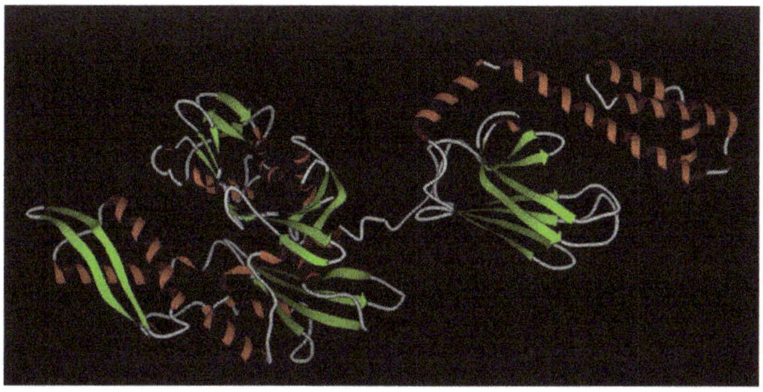

Hsp70 (pdb database 2KHO). Left domain is NBD (Nucleotide Binding domain) which can bind to ATP and ADP. Right lower domain is the groove, Subtrate Binding domain. Right upper domain is the "lid". (α-helices are shown in red. β-strands are shown in green.)

The Hsp70 proteins have three major functional domains:

- **N-terminal ATPase domain** – binds ATP (Adenosine triphosphate) and hydrolyzes it to ADP (Adenosine diphosphate). The NBD consists of two lobes with a deep cleft between them, at the bottom of which nucleotide (ATP and ADP) binds. The exchange of ATP and ADP leads to conformational changes in the other two domains.
- **Substrate binding domain** – is composed of a 25 kDa β sheet subdomain and a 10 kDa helical subdomain. The β sheet subdomain consists of stranded β sheets with upward protruding loops, as a typical β barrel, which enclose the peptide backbone of the substrate. SBD contains a groove with an affinity for neutral, hydrophobic amino acid residues. The groove is long enough to interact with peptides up to seven residues in length.
- **C-terminal domain** – rich in alpha helical structure acts as a 'lid' for the substrate binding domain. The helical subdomain consists of five helices, with two helices packed against two sides of the β sheet subdomain, stabilizing the inner structure. In addition, one of the helix forms a salt bridge and several hydrogen bonds to the outer Loops, thereby closing the substrate-binding pocket like a lid. Three helices in this domain form another hydrophobic core which may be stabilization of the "lid". When an Hsp70 protein is ATP bound, the lid is open and peptides bind and release relatively rapidly. When Hsp70 proteins

are ADP bound, the lid is closed, and peptides are tightly bound to the substrate binding domain.

Function and Regulation

The Hsp70 system interacts with extended peptide segments of proteins as well as partially folded proteins to prevent aggregation, remodel folding pathways, and regulate activity When not interacting with a substrate peptide, Hsp70 is usually in an ATP bound state. Hsp70 by itself is characterized by a very weak ATPase activity, such that spontaneous hydrolysis will not occur for many minutes. As newly synthesized proteins emerge from the ribosomes, the substrate binding domain of Hsp70 recognizes sequences of hydrophobic amino acid residues, and interacts with them. This spontaneous interaction is reversible, and in the ATP bound state Hsp70 may relatively freely bind and release peptides. However, the presence of a peptide in the binding domain stimulates the ATPase activity of Hsp70, increasing its normally slow rate of ATP hydrolysis. When ATP is hydrolyzed to ADP the binding pocket of Hsp70 closes, tightly binding the now-trapped peptide chain. Further speeding ATP hydrolysis are the so-called J-domain cochaperones: primarily Hsp40 in eukaryotes, and DnaJ in prokaryotes. These cochaperones dramatically increase the ATPase activity of Hsp70 in the presence of interacting peptides.

By binding tightly to partially synthesized peptide sequences (incomplete proteins), Hsp70 prevents them from aggregating and being rendered nonfunctional. Once the entire protein is synthesized, a nucleotide exchange factor (BAG-1 and HspBP1 are among those which have been identified) stimulates the release of ADP and binding of fresh ATP, opening the binding pocket. The protein is then free to fold on its own, or to be transferred to other chaperones for further processing. HOP (the **H**sp70/Hsp90 **O**rganizing **P**rotein) can bind to both Hsp70 and Hsp90 at the same time, and mediates the transfer of peptides from Hsp70 to Hsp90.

Hsp70 also aids in transmembrane transport of proteins, by stabilizing them in a partially folded state. It is also known to be phosphorylated which regulates several of its functions.

Hsp70 proteins can act to protect cells from thermal or oxidative stress. These stresses normally act to damage proteins, causing partial unfolding and possible aggregation. By temporarily binding to hydrophobic residues exposed by stress, Hsp70 prevents these partially denatured proteins from aggregating, and allows them to refold. Low ATP is characteristic of heat shock and sustained binding is seen as aggregation suppression, while recovery from heat shock involves substrate binding and nucleotide cycling. In a thermophile anaerobe (*Thermotoga maritima*) the Hsp70 demonstrates redox sensitive binding to model peptides, suggesting a second mode of binding regulation based on oxidative stress.

Hsp70 seems to be able to participate in disposal of damaged or defective proteins. Interaction with CHIP (*C*arboxyl-terminus of *H*sp70 *I*nteracting *P*rotein)–an E3 ubiquitin ligase–allows Hsp70 to pass proteins to the cell's ubiquitination and proteolysis pathways.

Finally, in addition to improving overall protein integrity, Hsp70 directly inhibits apoptosis. One hallmark of apoptosis is the release of cytochrome c, which then recruits

Apaf-1 and dATP/ATP into an apoptosome complex. This complex then cleaves procaspase-9, activating caspase-9 and eventually inducing apoptosis via caspase-3 activation. Hsp70 inhibits this process by blocking the recruitment of procaspase-9 to the Apaf-1/dATP/cytochrome c apoptosome complex. It does not bind directly to the procaspase-9 binding site, but likely induces a conformational change that renders procaspase-9 binding less favorable. Hsp70 is shown to interact with Endoplasmic reticulum stress sensor protein IRE1alpha thereby protecting the cells from ER stress - induced apoptosis. This interaction prolonged the splicing of XBP-1 mRNA thereby inducing transcriptional upregulation of targets of spliced XBP-1 like EDEM1, ERdj4 and P58IPK rescuing the cells from apoptosis. Other studies suggest that Hsp70 may play an anti-apoptotic role at other steps, but is not involved in Fas-ligand-mediated apoptosis (although Hsp 27 is). Therefore, Hsp70 not only saves important components of the cell (the proteins) but also directly saves the cell as a whole. Considering that stress-response proteins (like Hsp70) evolved before apoptotic machinery, Hsp70›s direct role in inhibiting apoptosis provides an interesting evolutionary picture of how more recent (apoptotic) machinery accommodated previous machinery (Hsps), thus aligning the improved integrity of a cell›s proteins with the improved chances of that particular cell›s survival.

Cancer

Hsp70 is overexpressed in malignant melanoma and underexpressed in renal cell cancer.

Expression in Skin Tissue

Both HSP70 and HSP47 were shown to be expressed in dermis and epidermis following laser irradiation, and the spatial and temporal changes in HSP expression patterns define the laser-induced thermal damage zone and the process of healing in tissues. HSP70 may define biochemically the thermal damage zone in which cells are targeted for destruction, and HSP47 may illustrate the process of recovery from thermally induced damage.

Family Members

Prokaryotes express three Hsp70 proteins: **DnaK**, **HscA (Hsc66)**, and **HscC (Hsc62)**.

Eukaryotic organisms express several slightly different Hsp70 proteins. All share the common domain structure, but each has a unique pattern of expression or subcellular localization. These are, among others:

- **Hsc70** (Hsp73/HSPA8) is a constitutively expressed chaperone protein. It typically makes up one to three percent of total cellular protein.
- **Hsp70** (encoded by three very closely related paralogs: HSPA1A, HSPA1B, and HSPA1L) is a stress-induced protein. High levels can be produced by cells in response to hyperthermia, oxidative stress, and changes in pH.
- Binding immunoglobulin protein (**BiP** or **Grp78**) is a protein localized to the endoplasmic reticulum. It is involved in protein folding there, and can be upregulated in response to stress or starvation.

- **mtHsp70** or **Grp75** is the mitochondrial Hsp70.

The following is a list of human Hsp70 genes and their corresponding proteins:

gene	protein	synonyms	subcellular location
HSPA1A	Hsp70	HSP70-1, Hsp72	Nuc/Cyto
HSPA1B	Hsp70	HSP70-2	Nuc/Cyto
HSPA1L	Hsp70		?
HSPA2	Hsp70-2		?
HSPA4	Hsp70-4		?
HSPA4L	Hsp70-4L		?
HSPA5	Hsp70-5	BiP/Grp78	ER
HSPA6	Hsp70-6		?
HSPA7	Hsp70-7		?
HSPA8	Hsp70-8	Hsc70	Nuc/Cyto
HSPA9	Hsp70-9	Grp75/mtHsp70	Mito
HSPA12A	Hsp70-12a		?
HSPA14	Hsp70-14		?

GATA1

GATA-1 is a member of the GATA transcription factor family and is a key mediator of the development of specific types of blood cells from their precursor cells, termed hematopoietic progenitors or precursors. This protein plays a role in erythroid development by regulating a large ensemble of genes that mediate both the development and function of red blood cells. Critical functions in the developing red blood cell (erythroblast) include the establishment of the erythroid cytoskeleton, enzymes that mediate heme biosynthesis, and polypeptide chains that constitute the hemoglobin tetramer. Mutations in the gene encoding GATA-1 have been associated with X-linked dyserythropoietic anemia and thrombocytopenia.

Function

GATA1 is required for the maturation of red blood cells, megakaryocytes, mast cells and eosinophils. GATA1 mutant mice die in early embryonic development due to inability to form mature erythroid cells. GATA1 mutation in humans causes congenital anemias and thrombocytopenias.

GATA1 was first described as a red blood cell lineage transcription factor that activates the beta-globin gene. During red blood cell maturation, GATA1 activates nearly all erythroid-specific genes while silencing genes associated with the immature proliferative red blood cell precursor cells (erythroblasts). Genome-wide studies have provided evidence that GATA1 activates and represses a large number of genes. Many questions remain unanswered regarding the function of a large number of genes. By contrast, other GATA-1 target genes have established activities to control fundamental cell biological functions, including machinery that controls the ability of erythroid precursor cells to proliferate and proteins that control the capacity of the erythroid precursor cell to remodel its organelles such as mitochondria, proteins that control the RNA content of the erythroid precursor cell, and proteins that control signal transduction networks that orchestrate the many dynamic transitions of the developing erythroid precursor.

Structure

The GATA-1 protein contains multiple functional domains including the C-finger, the N-finger, and N-terminal sequences that have been suggested to constitute a transcriptional activation domain. The C-finger, named for being near the C-terminal, mediates Zinc finger sequence-specific DNA binding. The primary function of the N-finger, named for being near the N-terminal is binding to a cofactor named FOG1 (friend of GATA), although it has also been implicated in binding to naked DNA (DNA studied in the test tube that is not assembled into chromatin). The gene for GATA1 is on the X-chromosome.

Disease Linkage

Mutations in GATA1 cause anemias and thrombocytopenia in human patients. Disease-causing GATA1 mutations are present in the zinc finger DNA binding domains as well as protein-protein interaction domains of GATA1.

Mutations in exon 2 of the GATA1 gene are present in almost all cases of Down syndrome (DS)-associated acute megakaryoblastic leukemia (AMKL). While AMKL is typically associated with the (1;22) translocation and expression of a mutant fusion protein, the genetic alterations that promote individuals with DS-AMKL are related to the GATA1 mutations, and the formation of a truncated transcription factor (GATA1s).

The same mutations in exon 2 of GATA1 present in almost all Down Syndrome-associated transient myeloproliferative disorder (TMD) or transient leukemia (TL), a precursor condition that evolves into AMKL in 30% of patients, that as many as 10% of Down Syndrome children may develop.The incidence for the GATA1 mutation in about 4% of Down Syndrome patients, but less than 10% of those with the mutation developed AMKL.This mutation is present in the fetus, suggesting an early role in leukemogenesis. In addition to screening for TL, a GATA1 mutation at birth might serve as a bio-marker for an increased risk of DS-related AMKL.

Transcription Factor II D

Transcription factor II D ($TF_{II}D$) is one of several general transcription factors that make up the RNA polymerase II preinitiation complex. RNA polymerase II holoenzyme is a form of eukaryotic RNA polymerase II that is recruited to the promoters of protein-coding genes in living cells. It consists of RNA polymerase II, a subset of general transcription factors, and regulatory proteins known as SRB proteins. Before the start of transcription, the transcription Factor II D (TFIID) complex binds to the TATA box in the core promoter of the gene.

Functions

- Coordinates the activities of more than 70 polypeptides required for initiation of transcription by RNA polymerase II
- Binds to the core promoter to position the polymerase properly
- Serves as the scaffold for assembly of the remainder of the transcription complex
- Acts as a channel for regulatory signals

Structure

$TF_{II}D$ is itself composed of several subunits called TATA-binding protein Associated Factors (**TBP-associated factors**, or **TAFs**). In a test tube, only TBP is necessary for transcription at promoters that contain a TATA box. TAFs, however, add promoter selectivity, especially if there is no TATA box sequence for TBP to bind to. TAFs are included in two distinct complexes, TFIID and B-TFIID. The TFIID complex is composed of TBP and more than eight TAFs. But, the majority of TBP is present in the B-TFIID complex, which is composed of TBP and TAFII170 (BTAF1) in a 1:1 ratio.TFIID and B-TFIID are not equivalent, since transcription reactions utilizing TFIID are responsive to gene specific transcription factors such as SP1, while reactions reconstituted with B-TFIID are not.

Subunits in the TFIID complex include:

- TBP (TATA binding protein)
- TAF1 (TAFII250)
- TAF2 (CIF150)
- TAF3 (TAFII140)
- TAF4 (TAFII130/135)
- TAF4B (TAFII105)
- TAF5 (TAFII100)
- TAF6 (TAFII70/80)
- TAF7 (TAFII55)
- TAF8 (TAFII43)
- TAF9 (TAFII31/32)
- TAF9B (TAFII31L)
- TAF10 (TAFII30)
- TAF11 (TAFII28)
- TAF12 (TAFII20/15)
- TAF13 (TAFII18)
- TAF15 (TAFII68)

EUKARYOTIC TRANSCRIPTION

Eukaryotic transcription is the elaborate process that eukaryotic cells use to copy genetic information stored in DNA into units of RNA replica. Gene transcription occurs in both eukaryotic and prokaryotic cells. Unlike prokaryotic RNA polymerase that initiates the transcription of all different types of RNA, RNA polymerase in eukaryotes (including humans) comes in three variations, each encoding a different type of gene. A eukaryotic cell has a nucleus that separates the processes of transcription and translation. Eukaryotic transcription occurs within the nucleus where DNA is packaged into nucleosomes and higher order chromatin structures. The complexity of the eukaryotic genome necessitates a great variety and complexity of gene expression control.

Transcription is the process of copying genetic information stored in a DNA strand into a transportable complementary strand of RNA. Eukaryotic transcription takes place in the nucleus of the cell and proceeds in three sequential stages: initiation, elongation, and termination. The transcriptional machinery that catalyzes this complex reaction has at its core three multi-subunit RNA polymerases. RNA polymerase I is responsible for transcribing RNA that codes for genes that become structural components of the ribosome.

Protein coding genes are transcribed into messenger RNAs (mRNAs) that carry the information from DNA to the site of protein synthesis.Although mRNAs possess great diversity, they are not the most abundant RNA species made in the cell. The so-called non-coding RNAs account for the large majority of the transcriptional output of a cell. These non-coding RNAs perform a variety of important cellular functions.

RNA Polymerase

Eukaryotes have three nuclear RNA polymerases, each with distinct roles and properties

Name	Location	Product
RNA Polymerase I (Pol I, Pol A)	nucleolus	larger ribosomal RNA (rRNA) (28S, 18S, 5.8S)
RNA Polymerase II (Pol II, Pol B)	nucleus	messenger RNA (mRNA), most small nuclear RNAs (snRNAs), small interfering RNA (siRNAs) and micro RNA (miRNA).
RNA Polymerase III (Pol III, Pol C)	nucleus (and possibly the nucleolus-nucleoplasm interface)	transfer RNA (tRNA), other small RNAs (including the small 5S ribosomal RNA (5s rRNA), snRNA U6, signal recognition particle RNA (SRP RNA) and other stable short RNAs

RNA polymerase I (Pol I) catalyses the transcription of all rRNA genes except 5S. These rRNA genes are organised into a single transcriptional unit and are transcribed into a continuous transcript. This precursor is then processed into three rRNAs: 18S, 5.8S, and 28S. The transcription of rRNA genes takes place in a specialised structure of the nucleus called the nucleolus, where the transcribed rRNAs are combined with proteins to form ribosomes.

RNA polymerase II (Pol II) is responsible for the transcription of all mRNAs, some snRNAs, siRNAs, and all miRNAs. Many Pol II transcripts exist transiently as single strand precursor RNAs (pre-RNAs) that are further processed to generate mature RNAs.

For example, precursor mRNAs (pre-mRNAs)are extensively processed before exiting into the cytoplasm through the nuclear pore for protein translation.

RNA polymerase III (Pol III) transcribes small non-coding RNAs, including tRNAs, 5S rRNA, U6 snRNA, SRP RNA, and other stable short RNAs such as ribonuclease P RNA.

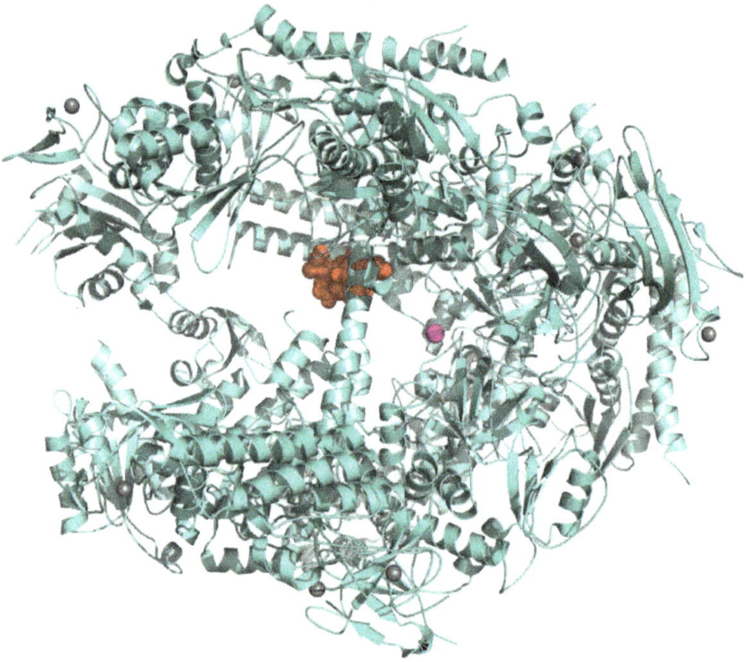

Structure of eukaryotic RNA polymerase II (light blue) in complex with α-amanitin (red), a strong poison found in death cap mushrooms that targets this vital enzyme.

RNA Polymerases I, II, and III contain 14, 12, and 17 subunits, respectively. All three eukaryotic polymerases have five core subunits that exhibit homology with the β, β', α^{I}, α^{II}, and ω subunits of E. coli RNA polymerase. An identical ω-like subunit (RBP6) is used by all three eukaryotic polymerases, while the same α-like subunits are used by Pol I and III. The three eukaryotic polymerases share four other common subunits among themselves. The remaining subunits are unique to each RNA polymerase. The additional subunits found in Pol I and Pol III relative to Pol II, are homologous to Pol II transcription factors.

Crystal structures of RNA polymerases I and II provide an opportunity to understand the interactions among the subunits and the molecular mechanism of eukaryotic transcription in atomic detail.

The carboxyl terminal domain (CTD) of RPB1, the largest subunit of RNA polymerase II, plays an important role in bringing together the machinery necessary for the synthesis and processing of Pol II transcripts.Long and structurally disordered, the CTD contains multiple repeats of heptapeptide sequence YSPTSPS that are subject to phosphorylation and other posttranslational modifications during the transcription cycle. These modifications and their regulation constitute the operational code for the CTD to control transcription initiation, elongation and termination and to couple transcription and RNA processing.

Initiation

The initiation of gene transcription in eukaryotes occurs in specific steps. First, an RNA polymerase along with general transcription factors binds to the promoter region of the gene to form a closed complex called the preinitiation complex. The subsequent transition of the complex from the closed state to the open state results in the melting or separation of the two DNA strands and the positioning of the template strand to the active site of the RNA polymerase. Without the need of a primer, RNA polymerase can initiate the synthesis of a new RNA chain using the template DNA strand to guide ribonucleotide selection and polymerization chemistry. However, many of the initiated syntheses are aborted before the transcripts reach a significant length (~10 nucleotides). During these abortive cycles, the polymerase keeps making and releasing short transcripts until it is able to produce a transcript that surpasses ten nucleotides in length. Once this threshold is attained, RNA polymerase escapes the promoter and transcription proceeds to the elongation phase.

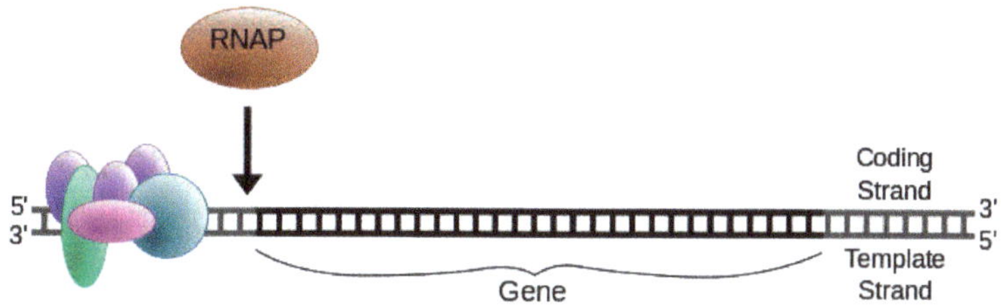

Here is a diagram of the attachment of RNA polymerase II to the de-helicized DNA. Credit: Forluvoft.

Eukaryotic Promoters and General Transcription Factors

Pol II-transcribed genes contain a region in the immediate vicinity of the transcription start site (TSS) that binds and positions the preinitiation complex. This region is called the core promoter because of its essential role in transcription initiation. Different classes of sequence elements are found in the promoters. For example, the TATA box is the highly conserved DNA recognition sequence for the TATA box binding protein, TBP, whose binding initiates transcription complex assembly at many genes.

Eukaryotic genes also contain regulatory sequences beyond the core promoter. These cis-acting control elements bind transcriptional activators or repressors to increase or decrease transcription from the core promoter. Well-characterized regulatory elements include enhancers, silencers, and insulators. These regulatory sequences can be spread over a large genomic distance, sometimes located hundreds of kilobases from the core promoters.

General transcription factors are a group of proteins involved in transcription initiation and regulation. These factors typically have DNA-binding domains that bind specific sequence elements of the core promoter and help recruit RNA polymerase to the transcriptional start site. General transcription factors for RNA polymerase II include TFIID, TFIIA, TFIIB, TFIIF, TFIIE, and TFIIH.

Assembly of Preinitiation Complex

To prepare for transcription, a complete set of general transcription factors and RNA polymerase need to be assembled at the core promoter to form the ~2 million dalton preinitiation complex. For example, for promoters that contain a TATA box near the TSS, the recognition of TATA box by the TBP subunit of TFIID initiates the assembly of a transcription complex. The next proteins to enter are TFIIA and TFIIB, which stabilize the DNA-TFIID complex and recruit Pol II in association with TFIIF and additional transcription factors. TFIIF serves as the bridge between the TATA-bound TBP and polymerase. One of the last transcription factors to be recruited to the preinitiation complex is TFIIH, which plays an important role in promoter melting and escape.

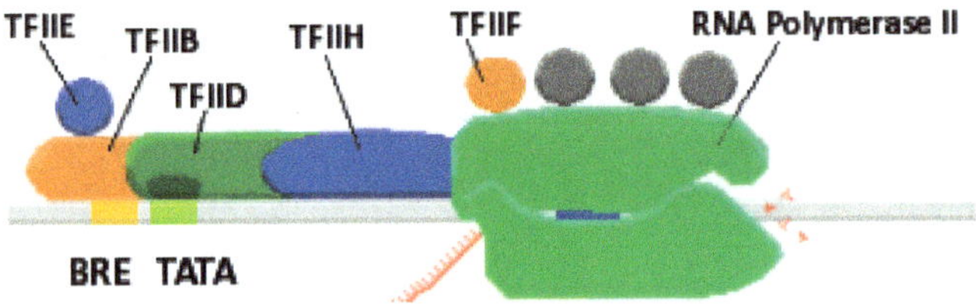

The diagram describes the eukaryotic preinitiation complex which includes the general transcription factors and RNA Polymerase II. Credit: ArneLH.

Promoter Melting and open Complex Formation

For pol II-transcribed genes, and unlike bacterial RNA polymerase, promoter melting requires hydrolysis of ATP and is mediated by TFIIH. TFIIH is a ten-subunit protein, including both ATPase and protein kinase activities. While the upstream promoter DNA is held in a fixed position by TFIID, TFIIH pulls downstream double-stranded DNA into the cleft of the polymerase, driving the separation of DNA strands and the transition of the preinitiation complex from the closed to open state. TFIIB aids in open complex formation by binding the melted DNA and stabilizing the transcription bubble.

Abortive Initiation

Once the initiation complex is open, the first ribonucleotide is brought into the active site to initiate the polymerization reaction in the absence of a primer. This generates a nascent RNA chain that forms a hetero-duplex with the template DNA strand. However, before entering the elongation phase, polymerase may terminate prematurely and release a short, truncated transcript. This process is called abortive initiation. Many cycles of abortive initiation may occur before the transcript grows to sufficient length to promote polymerase escape from the promoter. Throughout abortive initiation cycles, RNA polymerase remains bound to the promoter and pulls downstream DNA into its catalytic cleft in a scrunching-kind of motion.

Promoter Escape

When a transcript attains the threshold length of ten nucleotides, it enters the RNA exit channel. The polymerase breaks its interactions with the promoter elements and any regulatory proteins associated with the initiation complex that it no longer needs. Promoter escape in eukaryotes requires ATP hydrolysis and, in the case of Pol II-phosphorylation of the CTD. Meanwhile, the transcription bubble collapses down to 12-14 nucleotides, providing kinetic energy required for the escape.

Elongation

After escaping the promoter and shedding most of the transcription factors for initiation, the polymerase acquires new factors for the next phase of transcription: elongation. Transcription elongation is a processive process. Double stranded DNA that enters from the front of the enzyme is unzipped to avail the template strand for RNA synthesis. For every DNA base pair separated by the advancing polymerase, one hybrid RNA:DNA base pair is immediately formed. DNA strands and nascent RNA chain exit from separate channels; the two DNA strands reunite at the trailing end of the transcription bubble while the single strand RNA emerges alone.

Elongation Factors

Among the proteins recruited to polymerase are elongation factors, thus called because they stimulate transcription elongation. There are different classes of elongation factors. Some factors can increase the overall rate of transcribing, some can help the polymerase through transient pausing sites, and some can assist the polymerase to transcribe through chromatin. One of the elongation factors, P-TEFb, is particularly important. P-TEFb phosphorylates the second residue (Ser-2) of the CTD repeats (YSPTSPS) of the bound Pol II. P-TEFb also phosphorylates and activates SPT5 and TAT-SF1. SPT5 is a universal transcription factor that helps recruit 5'-capping enzyme to Pol II with a CTD phosphorylated at Ser-5. TAF-SF1 recruits components of the RNA splicing machinery to the Ser-2 phosphorylated CTD. P-TEFb also helps suppress transient pausing of polymerase when it encounters certain sequences immediately following initiation.

Transcription Fidelity

Transcription fidelity is achieved through multiple mechanisms. RNA polymerases select correct nucleoside triphosphate (NTP) substrate to prevent transcription errors. Only the NTP which correctly base pairs with the coding base in the DNA is admitted to the active center. RNA polymerase performs two known proof reading functions to detect and remove misincorporated nucleotides: pyrophosphorylytic editing and hydrolytic editing. The former removes the incorrectly inserted ribonucleotide by a simple reversal of the polymerization reaction, while the latter involves backtracking of the polymerase and cleaving of a segment of error-containing RNA product. Elongation factor TFIIS stimulates an inherent ribonuclease activity in the polymerase, allowing the removal of misincorporated bases through limited local RNA degradation. Note that all reactions (phosphodiester bond synthesis, pyrophosphorolysis, phosphodiester bond hydrolysis) are performed by RNA polymerase by using a single active center.

Pausing, Poising, and backtracking

Transcription elongation is not a smooth ride along the DNA railway. For proofreading, the polymerase is made to back-up, erase some of the RNA it has already made and have another go at transcription. In general, RNA polymerase does not transcribe through a gene at a constant pace. Rather it pauses periodically at certain sequences, sometimes for long periods of time before resuming transcription. In extreme cases, for example, when the polymerase encounters a damaged nucleotide, it comes to a complete halt. More often, an elongating polymerase is stalled near the promoter. Promoter-proximal pausing during early elongation is a commonly used mechanism for regulating genes poised to be expressed rapidly or in a coordinated fashion. Pausing is mediated by a complex called NELF (negative elongation factor) in collaboration with DSIF (DRB-sensitivity-inducing factor containing SPT4/SPT5). The blockage is released once the polymerase receives an activation signal, such as the phosphorylation of Ser-2 of CTD tail by P-TEFb. Other elongation factors such as ELL and TFIIS stimulate the rate of elongation by limiting the length of time that polymerase pauses.

RNA Processing

Elongating polymerase is associated with a set of protein factors required for various types of RNA processing. mRNA is capped as soon as it emerges from the RNA-exit channel of the polymerase. After capping, dephosphorylation of Ser-5 within the CTD repeats may be responsible for dissociation of the capping machinery. Further phosphorylation of Ser-2 causes recruitment of the RNA splicing machinery that catalyzes the removal of non-coding introns to generate mature mRNA. Alternative splicing expands the protein complements in eukaryotes. Just as with 5'-capping and splicing, the CTD tail is involved in recruiting enzymes responsible for 3'-polyadenylation, the final RNA processing event that is coupled with the termination of transcription.

Termination

The last stage of transcription is termination, which leads to the dissociation of the complete transcript and the release of RNA polymerase from the template DNA.The process differs for each of the three RNA polymerases. The mechanism of termination is the least understood of the three transcription stages.

Factor-Dependent Termination

The termination of transcription of pre-rRNA genes by polymerase Pol I is performed by a system that needs a specific transcription termination factor. The mechanism used bears some resemblance to the rho-dependent termination in prokaryotes. Eukaryotic cells contain hundreds of ribosomal DNA repeats, sometimes distributed over multiple chromosomes. Termination of transcription occurs in the ribosomal intergenic spacer region that contains several transcription termination sites upstream of a Pol I pausing site. Through a yet unknown mechanism, the 3'-end of the transcript is cleaved, generating a large primary rRNA molecule that is further processed into the mature 18S, 5.8S and 28S rRNAs.

As Pol II reaches the end of a gene, two protein complexes carried by the CTD, CPSF (cleavage and polyadenylation specificity factor) and CSTF (cleavage stimulation factor), recognize the poly-A signal in the transcribed RNA. Poly-A-bound CPSF and CSTF recruit other proteins to carry out RNA cleavage and then polyadenylation. Poly-A polymerase adds approximately 200 adenines to the cleaved 3' end of the RNA without a template. The long poly-A tail is unique to transcripts made by Pol II.

In the process of terminating transcription by Pol I and Pol II, the elongation complex does not dissolve immediately after the RNA is cleaved. The polymerase continues to move along the template, generating a second RNA molecule associated with the elongation complex. Two models have been proposed to explain how termination is achieved at last. The allosteric model states that when transcription proceeds through the termination sequence, it causes disassembly of elongation factors and/or an assembly of termination factors that cause conformational changes of the elongation complex. The torpedo model suggests that a 5' to 3' exonuclease degrades the second RNA as it emerges from the elongation complex. Polymerase is released as the highly processive exonuclease overtakes it. It is proposed that an emerging view will express a merge of these two models.

Factor-Independent Termination

RNA polymerase III can terminate transcription efficiently without involvement of additional factors. The Pol III termination signal consists of a stretch of thymines (on the nontemplate strand) located within 40bp downstream from the 3' end of mature RNAs. The poly-T termination signal pauses Pol III and causes it to back track to the nearest RNA hairpin to become a "dead-end" complex. Consistent with the allosteric mechanism of termination, the RNA hairpin allosterically opens Pol III and causes the elongation complex to disintegrate. The extensive structure embedded in the Pol III-transcript thus is responsible for the factor-independent release of Pol III at the end of a gene. RNA-duplex-dependent termination is an ancient mechanism that dates back to the last universal common ancestor.

Eukaryotic Transcriptional Control

The regulation of gene expression in eukaryotes is achieved through the interaction of several levels of control that acts both locally to turn on or off individual genes in response to a specific cellular need and globally to maintain a chromatin-wide gene expression pattern that shapes cell identity. Because eukaryotic genome is wrapped around histones to form nucelosomes and higher-order chromatin structures, the substrates for transcriptional machinery are in general partially concealed. Without regulatory proteins, many genes are expressed at low level or not expressed at all. Transcription requires displacement of the positioned nucleosomes to enable the transcriptional machinery to gain access of the DNA.

All steps in the transcription are subject to some degree of regulation. Transcription initiation in particular is the primary level at which gene expression is regulated. Targeting the rate-limiting initial step is the most efficient in terms of energy costs for the cell. Transcription initiation is regulated by cis-acting elements (enhancers, silencers, isolators) within the regulatory regions of the DNA, and sequence-specific trans-acting factors that act as activators or repressors. Gene transcription can also be regulated post-initiation by targeting the movement of the elongating polymerase.

Global Control and Epigenetic Regulation

The eukaryotic genome is organized into a compact chromatin structure that allows only regulated access to DNA. The chromatin structure can be globally "open" and more transcriptionally permissive or globally "condensed" and transcriptionally inactive. The former (euchromatin) is lightly packed and rich in genes under active transcription. The latter (heterochromatin) includes gene-poor regions such as telomeres and centromeres but also regions with normal gene density but transcriptionally silenced. Transcription can be silenced by histone modification (deaceltylation and methylation), RNA interference, and/or DNA methylation.

The gene expression patterns that define cell identity have to be inherited through cell division. This process is called epigenetic regulation. DNA methylation is reliably inherited through the action of maintenance methylases that modify the nascent DNA strand generated by replication. In mammalian cells, DNA methylation is the primary marker of transcriptionally silenced regions. Specialized proteins can recognize the marker and recruit histone deacetylases and methylases to re-establish the silencing. Nucleosome histone modifications could also be inherited during cell division, however, it is not clear whether it can work independently without the direction by DNA methylation.

Gene-Specific Activation

The two main tasks of transcription initiation are to provide RNA polymerase with an access to the promoter and to assemble general transcription factors with polymerase into a transcription initiation complex. Diverse mechanisms of initiating transcription by overriding inhibitory signals at the gene promoter have been identified. Eukaryotic genes have acquired extensive regulatory sequences that encompass a large number of regulator-binding sites and spread overall kilobases (sometimes hundreds of kilobases) from the promoter--both upstream and downstream.The regulator binding sites are often clustered together into units called enhancers. Enhancers can facilitate highly cooperative action of several transcription factors (which constitute enhanceosomes). Remote enhancers allow transcription regulation at a distance. Insulators situated between enhancers and promoters help define the genes that an enhancer can or cannot influence.

Eukaryotic transcriptional activators have separate DNA-binding and activating functions. Upon binding to its cis-element, an activator can recruit polymerase directly or recruit other factors needed by the transcriptional machinery. An activator can also recruit nucleosome modifiers that alter chromatin in the vicinity of the promoter and thereby help initiation. Multiple activators can work together, either by recruiting a common or two mutually dependent components of the transcriptional machinery, or by helping each other bind to their DNA sites. These interactions can synergize multiple signaling inputs and produce intricate transcriptional responses to address cellular needs.

Gene-specific repression

Eukaryotic transcription repressors share some of the mechanisms used by their prokaryotic counterparts. For example, by binding to a site on DNA that overlaps with the binding site of an activator, a repressor can inhibit binding of the activator. But more frequently, eukaryotic repressors inhibit the function of an activator by masking its activating domain, preventing its nuclear localization, promoting its degradation, or inactivating it through chemical modifications. Repressors can directly inhibit transcription initiation by binding to a site upstream of a promoter and interacting with the transcriptional machinery. Repressors can indirectly repress transcription by recruiting histone modifiers (deacetylases and methylases) or nucelosome remodeling enzymes that affect the accessibility of the DNA. Repressing histone and DNA modifications are also the basis of transcriptional silencing that can spread along the chromatin and switch off multiple genes.

Elongation and termination control

The elongation phase starts once assembly of the elongation complex has been completed, and progresses until a termination sequence is encountered. The post-initiation movement of RNA polymerase is the target of another class of important regulatory mechanisms. For example, the transcriptional activator Tat affects elongation rather than initiation during its regulation of HIV transcription. In fact, many eukaryotic genes are regulated by releasing a block to transcription elongation called promoter-proximal pausing. Pausing can influence chromatin structure at promoters to facilitate gene activity and lead to rapid or synchronous transcriptional responses when cells are exposed to an activation signal. Pausing is associated with the binding of two negative elongation factors, DSIF (SPT4/SPT5) and NELF, to the elongation complex. Other factors can also influence the stability and duration of the paused polymerase. Pause release is triggered by the recruitment of the P-TEFb kinase.

Transcription termination has also emerged as an important area of transcriptional regulation. Termination is coupled with the efficient recycling of polymerase. The factors associated with transcription termination can also mediate gene looping and thereby determine the efficiency of re-initiation.

Transcription-Coupled DNA Repair

When transcription is arrested by the presence of a lesion in the transcribed strand of a gene, DNA repair proteins are recruited to the stalled RNA polymerase to initiate a process called transcription-coupled repair. Central to this process is the general transcription factor TFIIH that has ATPase activity. TFIIH causes a conformational change in the polymerase, to expose the transcription bubble trapped inside, in order for the DNA repair enzymes to gain access to the lesion. Thus, RNA polymerase serves as damage-sensing protein in the cell to target repair enzymes to genes that are being actively transcribed.

Comparisons between prokaryotic and eukaryotic transcription

Eukaryotic transcription is more complex than prokaryotic transcription. For instance, in eukaryotes the genetic material (DNA), and therefore transcription, is primarily localized to the nucleus, where it is separated from the cytoplasm (in which translation occurs) by the nuclear membrane. This allows for the temporal regulation of gene expression through the sequestration of the RNA in the nucleus, and allows for selective transport of mature RNAs to the cytoplasm. Bacteria do not have a distinct nucleus that separates DNA from ribosome and mRNA is translated into protein as soon as it is transcribed. The coupling between the two processes provides an important mechanism for prokaryotic gene regulation.

At the level of initiation, RNA polymerase in prokaryotes (bacteria in particular) binds strongly to the promoter region and initiates a high basal rate of transcription. No ATP hydrolysis is needed for the close-to-open transition, promoter melting is driven by binding reactions that favor the melted conformation. Chromatin greatly impedes transcription in eukaryotes. Assembly of large multi-protein preinitiation complex is required for promoter-specific initiation. Promoter melting in eukaryotes requires hydrolysis of ATP. As a result, eukaryotic RNA polymerases exhibit a low basal rate of transcription initiation.

Regulation of Transcription in Cancer

In vertebrates, the majority of gene promoters contain a CpG island with numerous CpG sites. When many of a gene›s promoter CpG sites are methylated the gene becomes silenced. Colorectal cancers typically have 3 to 6 driver mutations and 33 to 66 hitchhiker or passenger mutations. However, transcriptional silencing may be of more importance than mutation in causing progression to cancer. For example, in colorectal cancers about 600 to 800 genes are transcriptionally silenced by CpG island methylation (see regulation of transcription in cancer). Transcriptional repression in cancer can also occur by other epigenetic mechanisms, such as altered expression of microRNAs. In breast cancer, transcriptional repression of BRCA1 may occur more frequently by over-expressed microRNA-182 than by hypermethylation of the BRCA1 promoter (see Low expression of BRCA1 in breast and ovarian cancers).

REFERENCES

1. Watson, James; Tania A. Baker; Stephen P. Bell; Alexander Gann; Michael Levine; Richard Losik; Stephen C. Harrison. Molecular Biology of the Gene

(7th ed.). Benjamin-Cummings Publishing Company. ISBN 978-0-321-76243-6.

2. Mattick, J. S. (1 November 2001). "Non-coding RNAs: the architects of eukaryotic complexity". EMBO Reports. 2 (11): 986–991. doi:10.1093/embo-reports/kve230. PMC 1084129Freely accessible. PMID 11713189.
3. Lodish Harvey, ...[et al]. Molecular cell biology (7th ed.). New York: W.H. Freeman and Co. ISBN 9781429234139.
4. Cramer P, Armache KJ, Baumli S, Benkert S, Brueckner F, Buchen C, Damsma GE, Dengl S, Geiger SR, Jasiak AJ, Jawhari A, Jennebach S, Kamenski T, Kettenberger H, Kuhn CD, Lehmann E, Leike K, Sydow JF, Vannini A (2008). "Structure of eukaryotic RNA polymerases". Annu Rev Biophys. 37: 337–52. doi:10.1146/annurev.biophys.37.032807.130008. PMID 18573085.
5. Sirri, Valentina; Silvio Urcuqui-Inchima; Pascal Roussel; Danièle Hernandez-Verdun (2008). "Nucleolus: the fascinating nuclear body". Histochem Cell Biol. 129 (1): 13–31. doi:10.1007/s00418-007-0359-6. PMC 2137947Freely accessible. PMID 18046571.
6. Fromont-Racine, Micheline; Senger, Bruno; Saveanu, Cosmin; Fasiolo, Franco (August 2003). "Ribosome assembly in eukaryotes". Gene. 313: 17–42. doi:10.1016/S0378-1119(03)00629-2.
7. Dieci, Giorgio; Fiorino, Gloria; Castelnuovo, Manuele; Teichmann, Martin; Pagano, Aldo (December 2007). "The expanding RNA polymerase III transcriptome". Trends in Genetics. 23 (12): 614–622. doi:10.1016/j.tig.2007.09.001. PMID 17977614.
8. Carter, R.; Drouin, G. (21 December 2009). "The Increase in the Number of Subunits in Eukaryotic RNA Polymerase III Relative to RNA Polymerase II Is due to the Permanent Recruitment of General Transcription Factors". Molecular Biology and Evolution. 27 (5): 1035–1043. doi:10.1093/molbev/msp316.
9. Fernández-Tornero, Carlos; Moreno-Morcillo, María; Rashid, Umar J.; Taylor, Nicholas M. I.; Ruiz, Federico M.; Gruene, Tim; Legrand, Pierre; Steuerwald, Ulrich; Müller, Christoph W. (23 October 2013). "Crystal structure of the 14-subunit RNA polymerase I". Nature. 502 (7473): 644–649. doi:10.1038/nature12636.
10. Cramer, P. (19 April 2001). "Structural Basis of Transcription: RNA Polymerase II at 2.8 Angstrom Resolution". Science. 292 (5523): 1863–1876. doi:10.1126/science.1059493.
11. Corden, Jeffry L. (13 November 2013). "RNA Polymerase II C-Terminal Domain: Tethering Transcription to Transcript and Template". Chemical Reviews. 113 (11): 8423–8455. doi:10.1021/cr400158h.
12. Butler, Jennifer; James T. Kadonaga (2002). "The RNA polymerase II core promoter: a key component in the regulation of gene expression". Genes & Dev. 16 (20): 2583–2592. doi:10.1101/gad.1026202. PMID 12381658.
13. Lenhard, Boris; Sandelin, Albin; Carninci, Piero (6 March 2012). "Metazoan promoters: emerging characteristics and insights into transcriptional regulation". Nature Reviews Genetics. 13: 233–45. doi:10.1038/nrg3163. PMID 22392219.

14. Orphanides, G; Lagrange, T; Reinberg, D (1 November 1996). "The general transcription factors of RNA polymerase II.". Genes & Development. 10 (21): 2657–2683. doi:10.1101/gad.10.21.2657. PMID 8946909.
15. ROEDER, R (September 1996). "The role of general initiation factors in transcription by RNA polymerase II". Trends in Biochemical Sciences. 21 (9): 327–335. doi:10.1016/S0968-0004(96)10050-5.
16. He, Yuan; Jie Fang; Dylan J. Taatjes; Eva Nogales (28 March 2013). "Structural visualization of key steps in human transcription initiation". Nature. 495: 481–486. doi:10.1038/nature11991. PMC 3612373Freely accessible. PMID 23446344.
17. Holstege, F. C.P. (15 December 1997). "Three transitions in the RNA polymerase II transcription complex during initiation". The EMBO Journal. 16 (24): 7468–7480. doi:10.1093/emboj/16.24.7468. PMC 1170346Freely accessible. PMID 9405375.
18. SVEJSTRUP, J; VICHI, P; EGLY, J (September 1996). "The multiple roles of transcription/repair factor TFIIH". Trends in Biochemical Sciences. 21 (9): 346–350. doi:10.1016/S0968-0004(96)10046-3. PMID 8870499.
19. Revyakin A, Liu C, Ebright RH, Strick TR (17 November 2006). "Abortive Initiation and Productive Initiation by RNA Polymerase Involve DNA Scrunching". Science. 314 (5802): 1139–1143. doi:10.1126/science.1131398. PMC 2754787Freely accessible. PMID 17110577.
20. Dvir, Arik (September 2002). "Promoter escape by RNA polymerase II". Biochimica et Biophysica Acta (BBA) - Gene Structure and Expression. 1577 (2): 208–223. doi:10.1016/S0167-4781(02)00453-0. PMID 12213653.
21. Pokholok, Dmitry K; Hannett, Nancy M; Young, Richard A (April 2002). "Exchange of RNA Polymerase II Initiation and Elongation Factors during Gene Expression In Vivo". Molecular Cell. 9 (4): 799–809. doi:10.1016/S1097-2765(02)00502-6.
22. Wade, Joseph T; Struhl, Kevin (April 2008). "The transition from transcriptional initiation to elongation". Current Opinion in Genetics & Development. 18 (2): 130–136. doi:10.1016/j.gde.2007.12.008.
23. Saunders, Abbie; Core, Leighton J.; Lis, John T. (August 2006). "Breaking barriers to transcription elongation". Nature Reviews Molecular Cell Biology. 7 (8): 557–567. doi:10.1038/nrm1981.
24. Arndt, Karen; Caroline Kane (1 October 2003). "Running with RNA polymerase: eukaryotic transcript elongation". Trends in Genetics. 19 (10): 543–550. doi:10.1016/j.tig.2003.08.008.
25. Brès, Vanessa; Yoh, Sunnie M; Jones, Katherine A (June 2008). "The multi-tasking P-TEFb complex". Current Opinion in Cell Biology. 20 (3): 334–340. doi:10.1016/j.ceb.2008.04.008.
26. Westover, Kenneth D.; Bushnell, David A.; Kornberg, Roger D. (November 2004). "Structural Basis of Transcription". Cell. 119 (4): 481–489. doi:10.1016/j.cell.2004.10.016.

27. Wang, Dong; Bushnell, David A.; Westover, Kenneth D.; Kaplan, Craig D.; Kornberg, Roger D. (December 2006). "Structural Basis of Transcription: Role of the Trigger Loop in Substrate Specificity and Catalysis". Cell. 127 (5): 941–954. doi:10.1016/j.cell.2006.11.023.

28. Wang, D.; Bushnell, D. A.; Huang, X.; Westover, K. D.; Levitt, M.; Kornberg, R. D. (28 May 2009). "Structural Basis of Transcription: Backtracked RNA Polymerase II at 3.4 Angstrom Resolution". Science. 324 (5931): 1203–1206. doi:10.1126/science.1168729.

29. Sosunov, Vasily; Ekaterina Sosunova; Arkady Mustaev; Irina Bass; Vadim Nikiforov; Alex Goldfarb (2003). "Unified two-metal mechanism of RNA synthesis and degradation by RNA polymerase". The EMBO Journal. 22: 2234–2244. doi:10.1093/emboj/cdg193.

30. Adelman, Karen; Lis, John T. (18 September 2012). "Promoter-proximal pausing of RNA polymerase II: emerging roles in metazoans". Nature Reviews Genetics. 13 (10): 720–731. doi:10.1038/nrg3293.

31. Missra, A.; Gilmour, D. S. (4 June 2010). "Interactions between DSIF (DRB sensitivity inducing factor), NELF (negative elongation factor), and the Drosophila RNA polymerase II transcription elongation complex". Proceedings of the National Academy of Sciences. 107 (25): 11301–11306. doi:10.1073/pnas.1000681107.

32. Richard, P.; Manley, J. L. (1 June 2009). "Transcription termination by nuclear RNA polymerases". Genes & Development. 23 (11): 1247–1269. doi:10.1101/gad.1792809.

33. Clancey, Suzanne (2008). "DNA transcription". Nature Education. 1 (41). Retrieved 27 November 2013.

34. Rosonina, E. (1 May 2006). "Terminating the transcript: breaking up is hard to do". Genes & Development. 20 (9): 1050–1056. doi:10.1101/gad.1431606.

35. Nielsen, S.; Yuzenkova, Y.; Zenkin, N. (27 June 2013). "Mechanism of Eukaryotic RNA Polymerase III Transcription Termination". Science. 340 (6140): 1577–1580. doi:10.1126/science.1237934.

36. Epshtein, Vitaly; Cardinale, Christopher J.; Ruckenstein, Andrei E.; Borukhov, Sergei; Nudler, Evgeny (December 2007). "An Allosteric Path to Transcription Termination". Molecular Cell. 28 (6): 991–1001. doi:10.1016/j.molcel.2007.10.011.

37. Shandilya, J; Robert SG (2012). "The transcription cycle in eukaryotes: from productive initiation to RNA polymerase II recycling". Biochim Biophys Acta. 1819 (5): 391–400. doi:10.1016/j.bbagrm.2012.01.010. PMID 22306664.

38. Kulaeva, Olga; Daria Gaykalova; Vasily M. Studitsky (2007). "Transcription Through Chromatin by RNA polymerase II: Histone Displacement and Exchange". Mutat Res. 618 (1–2): 116–129. doi:10.1016/j.mrfmmm.2006.05.040. PMC 1924643Freely accessible. PMID 17313961.

39. Peterlin, BM; DH Price (2006). "Controlling the elongation phase of transcription with P-TEFb". Molecular Cell. 23 (3): 297–305. doi:10.1016/j.molcel.2006.06.014. PMID 16885020.

40. Bannister, Andrew J; Kouzarides, Tony (15 February 2011). "Regulation of chromatin by histone modifications". Cell Research. 21 (3): 381–395. doi:10.1038/cr.2011.22. PMC 3193420Freely accessible. PMID 21321607.
41. Brown, Terrance A. (2002). Genomes. New York: Wiley-Liss. ISBN 0-471-25046-5.
42. Kao, Shaw-Yi; Andrew Calman; Paul Luciw; Matija Peterlin (3 December 1987). "Anti-termination of transcription within the long terminal repeat of HIV-1 by tat gene product". Nature. 330 (6147): 489–493. doi:10.1038/330489a0.
43. Lis, J (1998). "Promoter-associated Pausing in Promoter Architecture and Postinitiation Transcriptional Regulation". Cold Spring Harb Symp Quant Biol. 63: 347–56. doi:10.1101/sqb.1998.63.347. PMID 10384299.
44. Cheng B, Li T, Rahl PB, Adamson TE, Loudas NB, Guo J, Varzavand K, Cooper JJ, Hu X, Gnatt A, Young RA, Price DH (January 2012). "Functional Association of Gdown1 with RNA Polymerase II Poised on Human Genes". Molecular Cell. 45 (1): 38–50. doi:10.1016/j.molcel.2011.10.022. PMC 3259526Freely accessible. PMID 22244331.
45. Feige, MJ; LM Hendershot (April 2011). "Disulfide bonds in ER protein folding and homeostasis". Curr Opin Cell Biol. 23 (2): 167–75. doi:10.1016/j.ceb.2010.10.012. PMID 21144725.
46. MELLON, I (October 1987). "Selective removal of transcription-blocking DNA damage from the transcribed strand of the mammalian DHFR gene". Cell. 51 (2): 241–249. doi:10.1016/0092-8674(87)90151-6.
47. Sarker, Altaf H.; Tsutakawa, Susan E.; Kostek, Seth; Ng, Cliff; Shin, David S.; Peris, Marian; Campeau, Eric; Tainer, John A.; Nogales, Eva; Cooper, Priscilla K. (October 2005). "Recognition of RNA Polymerase II and Transcription Bubbles by XPG, CSB, and TFIIH: Insights for Transcription-Coupled Repair and Cockayne Syndrome". Molecular Cell. 20 (2): 187–198. doi:10.1016/j.molcel.2005.09.022. PMID 16246722.
48. Saxonov S, Berg P, Brutlag DL (2006). "A genome-wide analysis of CpG dinucleotides in the human genome distinguishes two distinct classes of promoters". Proc. Natl. Acad. Sci. U.S.A. 103 (5): 1412–7. doi:10.1073/pnas.0510310103. PMC 1345710Freely accessible. PMID 16432200.
49. Bird A (2002). "DNA methylation patterns and epigenetic memory". Genes Dev. 16 (1): 6–21. doi:10.1101/gad.947102. PMID 11782440.
50. Vogelstein B, Papadopoulos N, Velculescu VE, Zhou S, Diaz LA, Kinzler KW (2013). "Cancer genome landscapes". Science. 339 (6127): 1546–58. doi:10.1126/science.1235122. PMC 3749880Freely accessible. PMID 23539594.
51. Tessitore A, Cicciarelli G, Del Vecchio F, Gaggiano A, Verzella D, Fischietti M, Vecchiotti D, Capece D, Zazzeroni F, Alesse E (2014). "MicroRNAs in the DNA Damage/Repair Network and Cancer". Int J Genomics. 2014: 820248. doi:10.1155/2014/820248. PMC 3926391Freely accessible. PMID 24616890.

CHAPTER 6

Role of a Mitogen

Chapter Contents

A mitogen is any substance that triggers the process of mitosis, or cell division. Frequently, mitogens are proteins — compounds consisting of amino acid molecules that facilitate biochemical reactions in organisms. Mitogens are often encountered in the field of immunology due to their importance in inducing signal transduction pathways. These pathways are processes in an organism by which biochemical "signals" are passed from outside a cell to inside it, causing an alteration to take place within the cell. One such pathway, the mitogen-activated protein kinase (MAPK), can sometimes mutate and lead to cancer.

INTRODUCTION

A **mitogen** is a chemical substance that encourages a cell to commence cell division, triggering mitosis. A mitogen is usually some form of a protein. **Mitogenesis** is the induction (triggering) of mitosis, typically via a mitogen. Mitogens trigger signal transduction pathways in which mitogen-activated protein kinase (MAPK) is involved, leading to mitosis.

USE IN IMMUNOLOGY

Lymphocytes can enter mitosis when they are activated by mitogens or antigens. B cells specifically can divide when they encounter an antigen matching their immunoglobulin. T cells undergo mitosis when stimulated by mitogens to produce small lymphocytes that are then responsible for the production of lymphokines, which are substances that modify the host organism to improve its immunity. B cells, on the other hand, divide to produce plasma cells when stimulated by mitogens, which then produce immunoglobulins, or antibodies. Mitogens are often used to stimulate lymphocytes and thereby assess immune function. The most commonly used mitogens in clinical laboratory medicine are:

Name	Acts upon T cells?	Acts upon B cells?
phytohaemagglutinin (PHA)	yes	no
concanavalin A (conA)	yes	no
lipopolysaccharide (LPS)	no	yes
pokeweed mitogen (PWM)	yes	yes

Lipopolysaccharide toxin from gram-negative bacteria is thymus-independent. They may directly activate B cells, regardless of their antigenicspecificity. Plasma cells are terminally differentiated and, therefore, cannot undergo mitosis. Memory B cells can proliferate to produce more memory cells or plasma B cells. This is how the mitogen works, that is, by inducing mitosis in memory B cells to cause them to divide, with some becoming plasma cells.

Other uses

Mitogen-activated protein kinase (MAPK) pathways can induce enzymes such as the COX-2 enzyme. MAPK pathways may also play a role in the regulation of PTGS2.

GROWTH FACTOR

A **growth factor** is a naturally occurring substance capable of stimulating cellular growth, proliferation, healing, and cellular differentiation. Usually it is a protein or a steroid hormone. Growth factors are important for regulating a variety of cellular processes.

Growth factors typically act as signaling molecules between cells. Examples are cytokines and hormones that bind to specific receptors on the surface of their target cells.

They often promote cell differentiation and maturation, which varies between growth factors. For example, bone morphogenetic proteins stimulate bone cell differentiation, while fibroblast growth factors and vascular endothelial growth factors stimulate blood vessel differentiation (angiogenesis).

Growth Factors Versus Cytokines

Growth factor is sometimes used interchangeably among scientists with the term *cytokine*. Historically, cytokines were associated with hematopoietic (blood and lymph forming) cells and immune system cells (e.g., lymphocytes and tissue cells from spleen, thymus, and lymph nodes). For the circulatory system and bone marrow in which cells can occur in a liquid suspension and not bound up in solid tissue, it makes sense for them to communicate by soluble, circulating protein molecules. However, as different lines of research converged, it became clear that some of the same signaling proteins the hematopoietic and immune systems used were also being used by all sorts of other cells and tissues, during development and in the mature organism.

While *growth factor* implies a positive effect on cell division, *cytokine* is a neutral term with respect to whether a molecule affects proliferation. While some cytokines can be growth factors, such as G-CSF and GM-CSF, others have an inhibitory effect on cell growth or proliferation. Some cytokines, such as Fas ligand, are used as «death» signals; they cause target cells to undergo programmed cell death or *apoptosis*.

The growth factor was first discovered by Rita Levi-Montalcini, which won her a Nobel prize.

Classes of Growth Factors

Individual growth factor proteins tend to occur as members of larger families of structurally and evolutionarily related proteins. There are many families, some of which are listed below:

- Adrenomedullin (AM)
- Angiopoietin (Ang)
- Autocrine motility factor
- Bone morphogenetic proteins (BMPs)
- Ciliary neurotrophic factor family
  - Ciliary neurotrophic factor (CNTF)
  - Leukemia inhibitory factor (LIF)
  - Interleukin-6 (IL-6)
- Colony-stimulating factors
  - Macrophage colony-stimulating factor (m-CSF)
  - Granulocyte colony-stimulating factor (G-CSF)
  - Granulocyte macrophage colony-stimulating factor (GM-CSF)
- Epidermal growth factor (EGF)
- Ephrins
  - Ephrin A1
  - Ephrin A2
  - Ephrin A3
  - Ephrin A4
  - Ephrin A5
  - Ephrin B1
  - Ephrin B2

  - Ephrin B3
- Erythropoietin (EPO)
- Fibroblast growth factor (FGF)
- Foetal Bovine Somatotrophin (FBS)
- GDNF family of ligands
  - Glial cell line-derived neurotrophic factor (GDNF)
  - Neurturin
  - Persephin
  - Artemin
- Growth differentiation factor-9 (GDF9)
- Hepatocyte growth factor (HGF)
- Hepatoma-derived growth factor (HDGF)
- Insulin
- Insulin-like growth factors
  - Insulin-like growth factor-1 (IGF-1)
  - Insulin-like growth factor-2 (IGF-2)
- Interleukins
  - IL-1- Cofactor for IL-3 and IL-6. Activates T cells.
  - IL-2 – T-cell growth factor. Stimulates IL-1 synthesis. Activates B-cells and NK cells.
  - IL-3 – Stimulates production of all non-lymphoid cells.
  - IL-4 – Growth factor for activated B cells, resting T cells, and mast cells.
  - IL-5 – Induces differentiation of activated B cells and eosinophils.
  - IL-6 – Stimulates Ig synthesis. Growth factor for plasma cells.
  - IL-7 – Growth factor for pre-B cells.
- Keratinocyte growth factor (KGF)
- Migration-stimulating factor (MSF)
- Macrophage-stimulating protein (MSP), also known as hepatocyte growth factor-like protein (HGFLP)
- Myostatin (GDF-8)
- Neuregulins
  - Neuregulin 1 (NRG1)
  - Neuregulin 2 (NRG2)
  - Neuregulin 3 (NRG3)

- Neuregulin 4 (NRG4)

- Neurotrophins
  - Brain-derived neurotrophic factor (BDNF)
  - Nerve growth factor (NGF)
  - Neurotrophin-3 (NT-3)
  - Neurotrophin-4 (NT-4)
- Placental growth factor (PGF)
- Platelet-derived growth factor (PDGF)
- Renalase (RNLS) – Anti-apoptotic survival factor
- T-cell growth factor (TCGF)
- Thrombopoietin (TPO)
- Transforming growth factors
  - Transforming growth factor alpha (TGF-α)
  - Transforming growth factor beta (TGF-β)
  - Tumor necrosis factor-alpha (TNF-α)
  - Vascular endothelial growth factor (VEGF)
  - Wnt Signaling Pathway

Growth Factors in Platelets

The alpha granules in blood platelets contain growth factors PDGF, IGF-1, EGF, and TGF-β which begin healing of wounds by attracting and activating macrophages, fibroblasts, and endothelial cells.

Uses in Medicine

For the last two decades, growth factors have been increasingly used in the treatment of hematologic and oncologic diseases and cardiovascular diseases such as:

- neutropenia
- myelodysplastic syndrome (MDS)
- leukemias
- aplastic anaemia
- bone marrow transplantation
- angiogenesis for cardiovascular diseases

LECTIN

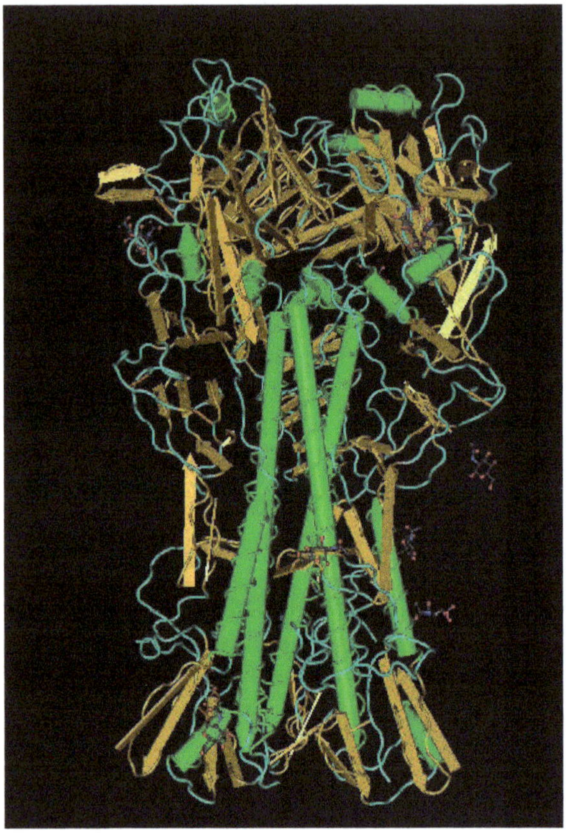

Lateral hemagglutinine.

Lectins are carbohydrate-binding proteins, macromolecules that are highly specific for sugar moieties.

Long before a deeper understanding of their numerous biological functions, the plant lectins, also known as phytohemagglutinins, were noted for their particular high specificity for foreign glycoconjugates (e.g. those of fungi, invertebrates, and animals). and used in biomedicine for blood cell testing and in biochemistry for fractionation.

Lectins perform recognition on the cellular and molecular level and play numerous roles in biological recognition phenomena involving cells, carbohydrates, and proteins. Lectins also mediate attachment and binding of bacteria and viruses to their intended targets. For example, it is hypothesized that some hepatitis C viral glycoproteins attach to C-type lectins on the host cell surface (liver cells) for infection.

Lectins may be disabled by specific mono- and oligosaccharides, which bind to ingested lectins from grains, legume, nightshade plants and dairy; binding can prevent their attachment to the carbohydrates within the cell membrane. Some lectins may be powerful toxins as for instance ricin, and others have been incorporated into genetically engineered crops to transfer traits, such as resistance to pests and resistance to herbicides.

Table of the major lectins			
Lectin Symbol	**Lectin name**	**Source**	**ligand motif**
Mannose binding lectins			
ConA	Concanavalin A	*Canavalia ensiformis*	α-D-mannosyl and α-D-glucosyl residues branched α-mannosidic structures (high α-mannose type, or hybrid type and biantennary complex type N-Glycans)
LCH	Lentil lectin	*Lens culinaris*	Fucosylated core region of bi- and triantennary complex type N-Glycans
GNA	Snowdrop lectin	*Galanthus nivalis*	α 1-3 and α 1-6 linked high mannose structures
Galactose / N-acetylgalactosamine binding lectins			
RCA	Ricin, Ricinus communis Agglutinin, RCA120	*Ricinus communis*	Galβ1-4GalNAcβ1-R
PNA	Peanut agglutinin	*Arachis hypogaea*	Galβ1-3GalNAcα1-Ser/Thr (T-Antigen)
AIL	Jacalin	*Artocarpus integrifolia*	(Sia)Galβ1-3GalNAcα1-Ser/Thr (T-Antigen)
VVL	Hairy vetch lectin	*Vicia villosa*	GalNAcα-Ser/Thr (Tn-Antigen)
N-acetylglucosamine binding lectins			
WGA	Wheat Germ Agglutinin, WGA	*Triticum vulgaris*	GlcNAcβ1-4GlcNAcβ1-4GlcNAc, Neu5Ac (sialic acid)
N-acetylneuraminic acid binding lectins			
SNA	Elderberry lectin	*Sambucus nigra*	Neu5Acα2-6Gal(NAc)-R
MAL´	Maackia amurensis leukoagglutinin	*Maackia amurensis*	Neu5Ac/Gcα2,3Galβ1,4Glc(NAc)
MAH	Maackia amurensis hemoagglutinin	*Maackia amurensis*	Neu5Ac/Gcα2,3Galβ1,3(Neu5Acα2,6)GalNac
Fucose binding lectins			
UEA	Ulex europaeus agglutinin	*Ulex europaeus*	Fucα1-2Gal-R
AAL	Aleuria aurantia lectin	*Aleuria aurantia*	Fucα1-2Galβ1-4(Fucα1-3/4)Galβ1-4GlcNAc, R2-GlcNAcβ1-4(Fucα1-6)GlcNAc-R1

The name "lectin" is derived from the Latin word *legere*, meaning, among other things, "to select".

History

Although they were first discovered more than 100 years ago in plants, now lectins are known to be present throughout nature. It is generally believed that the earliest description of a lectin was given by Peter Hermann Stillmark in his doctoral thesis presented in 1888 to the University of Dorpat. Stillmark isolated ricin, an extremely toxic hemagglutinin, from seeds of the castor plant (*Ricinus communis*).

The first lectin to be purified on a large scale and available on a commercial basis was concanavalin A, which is now the most-used lectin for characterization and purification of sugar-containing molecules and cellular structures. The legume lectins are probably the most well-studied lectins.

Biological Functions

Most lectins do not possess enzymatic activity. Lectins occur ubiquitously in nature. They may bind to a soluble carbohydrate or to a carbohydrate moiety that is a part of a glycoprotein or glycolipid. They typically agglutinate certain animal cells and/or precipitate glycoconjugates.

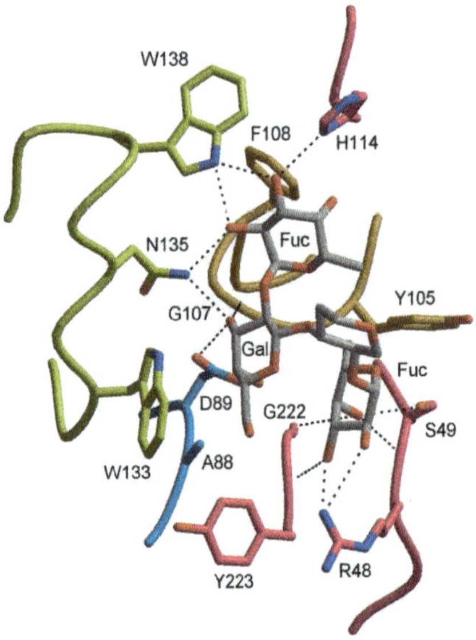

An oligosaccharide (shown in grey) bound in the binding site of a plant lectin (*Griffonia simplicifolia* isolectin IV in complex with the Lewis b blood group determinant); only a part of the oligosaccharide (central, in grey) is shown for clarity.

Functions in Animals

Lectins serve many different biological functions in animals, from the regulation of cell adhesion to glycoprotein synthesis and the control of protein levels in the blood. They also may bind soluble extracellular and intercellular glycoproteins. Some lectins are found on the surface of mammalian liver cells that specifically recognize galactose residues. It is believed that these cell-surface receptors are responsible for the removal of certain glycoproteins from the circulatory system.

Another lectin is a receptor that recognizes hydrolytic enzymes containing mannose-6-phosphate, and targets these proteins for delivery to the lysosomes. I-cell disease is one type of defect in this particular system.

Lectins also are known to play important roles in the immune system. Within the innate immune system lectins such as the MBL, the mannose-binding lectin, help mediate the first-line defense against invading microorganisms. Other lectins within the immune system are thought to play a role in self-nonself discrimination and they likely modulate inflammatory and autoreactive processes. Intelectins (X-type lectins) were shown to bind microbial glycans and may function in the innate immune system as well.

Functions in Plants

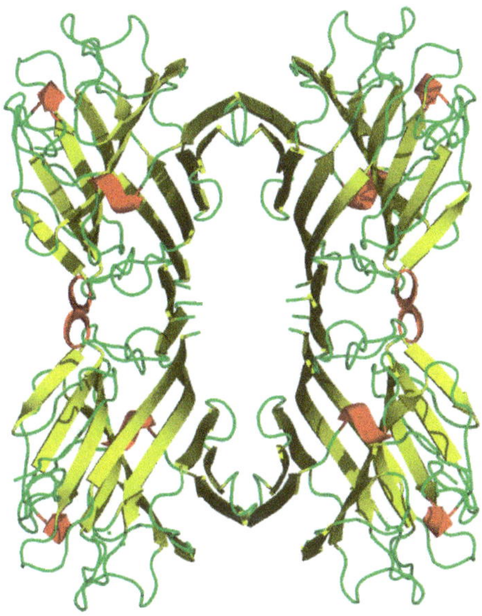

Leucoagglutinin, a toxic phytohemagglutinin found in raw *Vicia faba*.

The function of lectins in plants (legume lectin) is still uncertain. Once thought to be necessary for rhizobia binding, this proposed function was ruled out through lectin-knockout transgene studies.

The large concentration of lectins in plant seeds decreases with growth, and suggests a role in plant germination and perhaps in the seed's survival itself. The binding of glycoproteins on the surface of parasitic cells also is believed to be a function. Several plant lectins have been found to recognize non-carbohydrate ligands that are primarily hydrophobic in nature, including adenine, auxins, cytokinin, and indole acetic acid, as well as water-soluble porphyrins. It has been suggested that these interactions may be physiologically relevant, since some of these molecules function as phytohormones.

Use in Science, Medicine, and Technology

Use in Medicine and Medical Research

Purified lectins are important in a clinical setting because they are used for blood typing. Some of the glycolipids and glycoproteins on an individual's red blood cells can be identified by lectins.

- A lectin from *Dolichos biflorus* is used to identify cells that belong to the A1 blood group.
- A lectin from *Ulex europaeus* is used to identify the H blood group antigen.
- A lectin from *Vicia graminea* is used to identify the N blood group antigen.
- A lectin from *Iberis amara* is used to identify the M blood group antigen.
- A lectin from coconut milk is used to identify Theros antigen.
- A lectin from *Carex* is used to identify R antigen.

In neuroscience, the anterograde labeling method is used to trace the path of efferent axons with PHA-L, a lectin from the kidney bean.

A lectin (BanLec) from bananas inhibits HIV-1 in vitro. Achylectins, isolated from *Tachypleus tridentatus*, show specific agglutinating activity against human A-type erythrocytes. Anti-B agglutinins such as anti-BCJ and anti-BLD separated from *Charybdis japonica* and *Lymantria dispar*, respectively, are of value both in routine blood grouping and research.

Use in studying Carbohydrate Recognition by Proteins

Lectin histochemistry of fish muscles infected by a Myxozoa

Lectins from legume plants, such as PHA or concanavalin A, have been used widely as model systems to understand the molecular basis of how proteins recognize carbohydrates, because they are relatively easy to obtain and have a wide variety of sugar specificities. The many crystal structures of legume lectins have led to a detailed insight of the atomic interactions between carbohydrates and proteins.

Use as a Biochemical Tool

Concanavalin A and other commercially available lectins have been used widely in affinity chromatography for purifying glycoproteins.

In general, proteins may be characterized with respect to glycoforms and carbohydrate structure by means of affinity chromatography, blotting, affinity electrophoresis, and affinity immunoelectrophoreis with lectins as well as in microarrays, as in evanescent-field fluorescence-assisted lectin microarray.

Use in Biochemical Warfare

One example of the powerful biological attributes of lectins is the biochemical warfare agent ricin. The protein ricin is isolated from seeds of the castor oil plant and comprises two protein domains. Abrin from the jequirity pea is similar:

- One domain is a lectin that binds cell surface galactosyl residues and enables the protein to enter cells
- The second domain is an N-glycosidase that cleaves nucleobases from ribosomal RNA, resulting in inhibition of protein synthesis and cell death.

Toxicity

Digestion and Immune Distress

Foods with high concentrations of lectins, such as beans, cereal grains, seeds, nuts, and potatoes, may be harmful if consumed in excess in uncooked or improperly-cooked form. Adverse effects may include nutritional deficiencies, and immune (allergic) reactions. Possibly, most effects of lectins are due to gastrointestinal distress through interaction of the lectins with the epithelial cells of the gut. A recent in vitro study has suggested that the mechanism of lectin damage may occur by interfering with the repair of already-damaged epithelial cells.

Lectins are considered another major family of protein antinutrients (ANCs), which are specific sugar-binding proteins exhibiting reversible carbohydrate-binding activities. Lectins are similar to antibodies in their ability to agglutinate red blood cells.

The toxicity of lectins has been identified by consumption of food with high lectin content, which may lead to diarrhea, nausea, bloating, and vomiting.

Many legume seeds have been proven to contain high lectin activity, termed hemagglutinating activity. Soybean is the most important grain legume crop in this category. Its seeds contain high activity of soybean lectins (soybean agglutinin or SBA). SBA is able to disrupt small intestinal metabolism and damage small intestinal villi via the ability of lectins to bind with brush border surfaces in the distal part of small intestine.

Heat processing can reduce the toxicity of lectins, but low temperature or insufficient cooking may not completely eliminate their toxicity, as some plant lectins are resistant to heat.

Lectin and Leptin Resistance

Lectin may cause leptin resistance, affecting its functions (signal have high levels of leptin and several effects gathering to protect from lipid overload), as indicated by studies on effects of single nucleotide polymorphisms on the function of leptin and the leptin receptor. Such leptin resistance may translate into diseases, notably it could be responsible for obesity in humans who have high levels of leptin.

Lectin may involve in pattern recognition and pathogen elimination in the innate immunity of vertebrates including fishes.

CHAPTER 7

Mitotic Index

Chapter Contents

- **Introduction**
- **Purpose**
- **Calculation**
- **Formula**

Mitotic index is a measure for the proliferation status of a cell population. It is defined as the ratio between the number of cells in mitosis and the total number of cells. The percentage of cells undergoing mitosis or it is defined as the ratio of no. of cells in the dividing phase to the total number of cells observed. This will help to identify the region of most mitotic activities. Mitotic index help us to quantify the cell division. Mitotic index decreases with increasing distance from root tip. That means gradual decrease in cell division as move from the zone of cell division to the zone of cell elongation. The meristematic region in the root tip is the actively growing region and thus the mitotic index is high.

INTRODUCTION

Mitotic index is defined as the ratio between the number of cells in a population undergoing mitosis to the number of cells in a population not undergoing mitosis.

PURPOSE

The purpose of the mitotic index is to measure cellular proliferation. The mitotic index is an important prognostic factor predicting both overall survival and response to chemotherapy in most types of cancer. It may lose much of its predictive value for elderly populations. For example, a low mitotic index loses any prognostic value for women over 70 years old with breast cancer.

Additionally, the mitotic index is simply a measurement to determine the percentage of cells undergoing mitosis. Mitosis is the division of somatic cells when genetic information from one single cell is equally dispersed into two daughter cells. Durations of the cell cycle and mitosis vary in different cell types. An elevated mitotic index indicates more cells are dividing, and thus obvious in cancer cells, The mitotic index may be elevated during necessary processes to life, such as the normal growth of plants or animals, as well as cellular repair the site of an injury.

CALCULATION

Mitotic index can be calculated using the following operation : cells observed with visible chromosomes ÷ total number of cells visible.

A typical figure of mitotic index includes statements like 10 mitotic figures are noted per 10 high power fields followed by 4 mitotic figures noted per 50 high power fields. Mitotic index: 14 MF/10 400x fields

FORMULA

$$\text{Mitotic I} = \frac{(P+M+A+T)}{N} * 100\%$$

(P+M+A+T) — the sum of all cells in phase as prophase, metaphase, anaphase and telophase, respectively; **N** — total number of cells.

REFERENCES

1. Rudolph; et al. (1998). Correlation between mitotic and Ki-67 labeling indices in paraffin-embedded carcinoma specimens. Human Pathology. **29**: 1216–1222. doi:10.1016/s0046-8177(98)90248-9.
2. Baak, J. P. A.; Gudlaugsson, E.; Skaland, I.; Guo, L. H. R.; Klos, J.; Lende, T. H.; Søiland, H. V.; Janssen, E. A. M.; Zur Hausen, A. (2008). "Proliferation is the strongest prognosticator in node-negative breast cancer: Significance, error sources, alternatives and comparison with molecular prognostic markers". Breast Cancer Research and Treatment. **115** *(2): 241–254.* doi:10.1007/s10549-008-0126-y. PMID 18665447.
3. Urry; et al. (2014). Campbell Biology in Focus. Pearson.
4. Edexcel practical materials created by Salters-Nuffield Advanced Biology, copyright University of York Science Education Group

CHAPTER 8

Mitotic Catastrophe and Carcinogenesis

Chapter Contents

Mitotic catastrophe refers to a mechanism of delayed mitosis-linked cell death, a sequence of events resulting from premature or inappropriate entry of cells into mitosis that can be caused by chemical or physical stresses. Mitotic catastrophe is unrelated to programmed cell death or apoptosis and is observed in cells lacking functional apoptotic pathways. It is has been observed following delayed DNA damage induced by ionizing radiation. It can also be triggered by agents influencing the stability of microtubule spindles, various anticancer drugs and mitotic failure caused by defective cell cycle checkpoints. Mitotic catastrophe is the primary mechanism underlying reproductive cell death in cancer cells treated with ionizing radiation.

Not all cells die immediately following abnormal mitosis caused by mitotic catastrophe, but many do. Cells that do not immediately die are likely to create aneuploid cells following subsequent attempts at cell division posing a risk of oncogenesis (i.e. potentially leading to cancer). A very small fraction of these aneuploid cells produced by mitotic catastrophe might later reduce DNA ploidy by reductive division involving meiotic cell division pathways.

INTRODUCTION

Carcinogenesis or **oncogenesis** or **tumorigenesis** is the formation of a cancer, whereby normal cells are transformed into cancer cells. The process is characterized by changes at the cellular, genetic, and epigenetic levels and abnormal cell division, in some cancers forming a malignant mass.

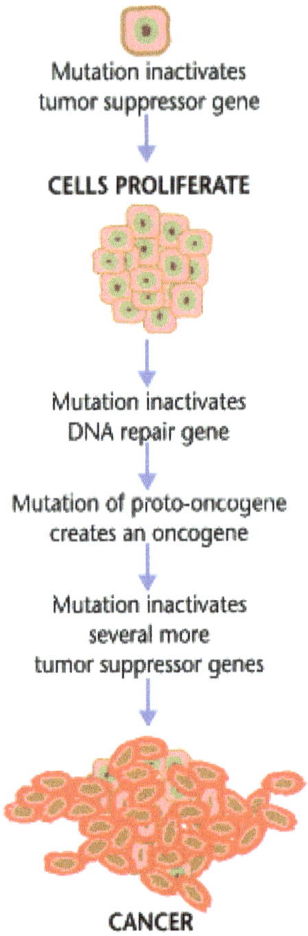

Cancers are caused by a series of mutations. Each mutation alters the behavior of the cell somewhat.

Cell division is a physiological process that occurs in almost all tissues and under many circumstances. Under normal circumstances, the balance between proliferation and programmed cell death, usually in the form of apoptosis, is maintained by regulation of both processes to ensure the integrity of tissues and organs. According to the prevailing accepted theory of carcinogenesis, the somatic mutation theory, mutations and epimutations in DNA that lead to cancer (only certain mutations and epimutations can lead to cancer and the majority of potential mutations and epimutations will have no such effect) disrupt these orderly processes by disrupting the programming regulating the processes, upsetting the normal balance between proliferation and cell death. This results in uncontrolled cell division and the evolution of those cells by natural selection in the body.

The uncontrolled and often rapid proliferation of cells can lead to benign tumors; some types of these may turn into malignant tumors (cancer). Benign tumors do not spread to other parts of the body or invade other tissues, and they are rarely a threat to life unless they compress vital structures or are physiologically active, for instance, producing a hormone. Malignant tumors can invade other organs, spread to distant locations (metastasis) and become life-threatening.

More than one mutation is necessary for carcinogenesis. In fact, a series of several mutations to certain classes of genes is usually required before a normal cell will transform into a cancer cell. On average, for example, 15 driver mutations and 60 passenger mutations are found in colon cancers. Mutations in those certain types of genes that play vital roles in cell division, apoptosis (cell death), and mutations and epimutations (see article Genome instability) in DNA repair genes will cause a cell to lose control of its cell proliferation.

Oncovirinae, viruses that contain an oncogene, are categorized as oncogenic because they trigger the growth of tumorous tissues in the host. This process is also referred to as viral transformation.

Cancer is fundamentally a disease of regulation of tissue growth. In order for a normal cell to transform into a cancer cell, genes that regulate cell growth and differentiation must be altered. Genetic and epigenetic changes can occur at many levels, from gain or loss of entire chromosomes, to a mutation affecting a single DNA nucleotide, or to silencing or activating a microRNA that controls expression of 100 to 500 genes. There are two broad categories of genes that are affected by these changes. Oncogenes may be normal genes that are expressed at inappropriately high levels, or altered genes that have novel properties. In either case, expression of these genes promotes the malignant phenotype of cancer cells. Tumor suppressor genes are genes that inhibit cell division, survival, or other properties of cancer cells. Tumor suppressor genes are often disabled by cancer-promoting genetic changes. Typically, changes in many genes are required to transform a normal cell into a cancer cell.

There is a diverse classification scheme for the various genomic changes that may contribute to the generation of cancer cells. Many of these changes are mutations, or changes in the nucleotide sequence of genomic DNA. There are also many epigenetic changes that alter whether genes are expressed or not expressed. Aneuploidy, the presence of an abnormal number of chromosomes, is one genomic change that is not a mutation, and may involve either gain or loss of one or more chromosomes through errors in mitosis.

Large-scale mutations involve the deletion or gain of a portion of a chromosome. Genomic amplification occurs when a cell gains many copies (often 20 or more) of a small chromosomal region, usually containing one or more oncogenes and adjacent genetic material. Translocation occurs when two separate chromosomal regions become abnormally fused, often at a characteristic location. A well-known example of this is the Philadelphia chromosome, or translocation of chromosomes 9 and 22, which occurs in chronic myelogenous leukemia, and results in production of the BCR-abl fusion protein, an oncogenic tyrosine kinase.

Small-scale mutations include point mutations, deletions, and insertions, which may occur in the promoter of a gene and affect its expression, or may occur in the gene›s coding sequence and alter the function or stability of its protein product. Disruption of a single gene may also result from integration of genomic material from a DNA virus or retrovirus, and such an event may also result in the expression of viral oncogenes in the affected cell and its descendants.

Epimutations include methylations or demethylations of the CpG islands of the promoter regions of genes, which result in repression or de-repression, respectively of gene expression. Epimutations can also occur by acetylation, methylation, phosphorylation or other alterations to histones, creating a histone code that represses or activates gene expression, and such histone epimutations can be important epigenetic factors in cancer. In addition, carcinogenic epimutation can occur through alterations of chromosome architecture caused by proteins such as HMGA2. A further source of epimutation is due to increased or decreased expression of microRNAs (miRNAs). For example, extra expression of miR-137 can cause downregulation of expression of 491 genes, and miR-137 is epigenetically silenced in 32% of colorectal cancers>

The cells in which all these DNA alterations accumulate are difficult to trace, but two recent lines of evidence suggest that normal stem cells may be the cells of origin in cancers. First, there exists a highly positive correlation (Spearman's rho = 0.81; $P < 3.5 \times$ 10−8) between the risk of developing cancer in a tissue and the number of normal stem cell divisions taking place in that same tissue. The correlation applied to 31 cancer types and extended across five orders of magnitude. This correlation means that if the normal stem cells from a tissue divide once, the cancer risk in that tissue is approximately 1X. If they divide 1,000 times, the cancer risk is 1,000X. And if the normal stem cells from a tissue divide 100,000 times, the cancer risk in that tissue is approximately 100,000X. This strongly suggests that the main reason we have cancer is that our normal stem cells divide, which implies that cancer originates in normal stem cells. Second, statistics show that most human cancers are diagnosed in aged people. A possible explanation is that cancers occur because cells accumulate damage through time. DNA is the only cellular component that can accumulate damage over the entire course of a life, and stem cells are the only cells that can transmit DNA from the zygote to cells late in life. Other cells cannot keep DNA from the beginning of life until a possible cancer occurs. This implies that most cancers arise from normal stem cells.

CAUSES

DNA Damage and Deficient DNA Repair in Carcinogenesis

DNA damage is considered to be the primary cause of cancer.More than 60,000 new naturally occurring DNA damages arise, on average, per human cell, per day, due to endogenous cellular processes (see article DNA damage (naturally occurring)).

Additional DNA damages can arise from exposure to exogenous agents. As one example of an exogenous carcinogeneic agent, tobacco smoke causes increased DNA damage, and these DNA damages likely cause the increase of lung cancer due to smoking. In other examples, UV light from solar radiation causes DNA damage that is important in melanoma, helicobacter pylori infection produces high levels of reactive oxygen species that damage DNA and contributes to gastric cancer, and the Aspergillus metabolite, aflatoxin, is a DNA damaging agent that is causative in liver cancer.

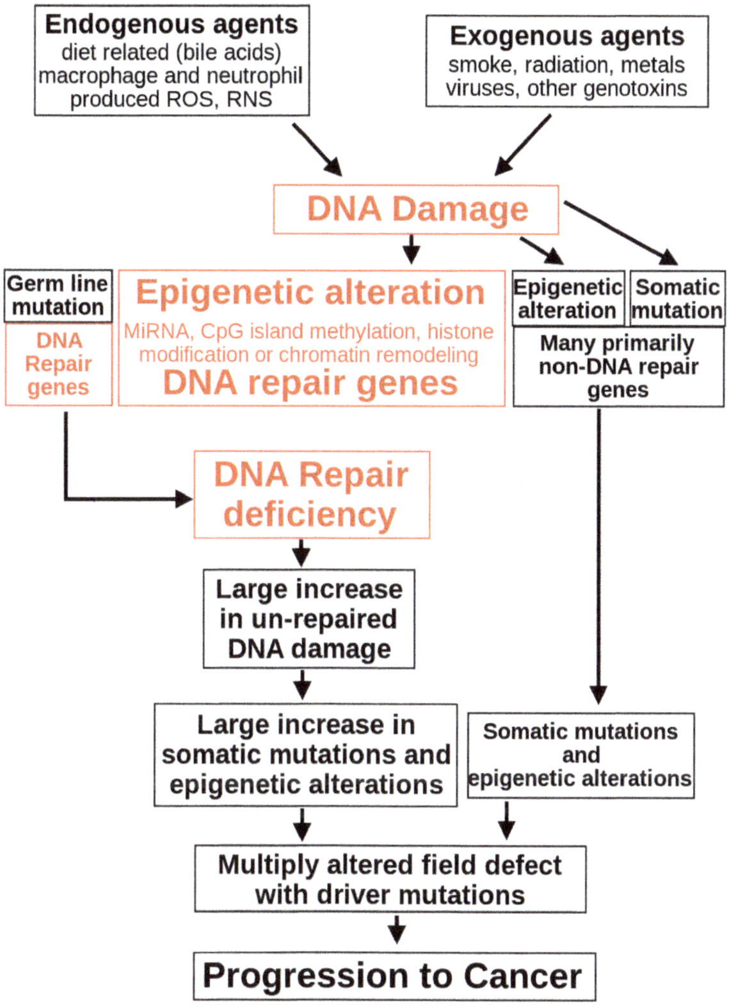

The central role of DNA damage and epigenetic defects in DNA repair genes in carcinogenesis.

DNA damages can also be caused by endogenous (naturally occurring) agents. Katsurano et al. indicated that macrophages and neutrophils in an inflamed colonic epithelium are the source of reactive oxygen species causing the DNA damages that initiate colonic tumorigenesis, and bile acids, at high levels in the colons of humans eating a high fat diet, also cause DNA damage and contribute to colon cancer.

Such exogenous and endogenous sources of DNA damage are indicated in the boxes at the top of the figure in this section. The central role of DNA damage in progression to cancer is indicated at the second level of the figure. The central elements of DNA damage, epigenetic alterations and deficient DNA repair in progression to cancer are shown in red.

A deficiency in DNA repair would cause more DNA damages to accumulate, and increase the risk for cancer. For example, individuals with an inherited impairment in any of 34 DNA repair genes (see article DNA repair-deficiency disorder) are at increased risk of cancer with some defects causing up to 100% lifetime chance of cancer (e.g. p53 mutations). Such germ line mutations are shown in a box at the left of the figure, with an indication of their contribution to DNA repair deficiency. However, such germline mutations (which cause highly penetrant cancer syndromes) are the cause of only about 1 percent of cancers.

The majority of cancers are called non-hereditary or sporadic cancers. About 30% of sporadic cancers do have some hereditary component that is currently undefined, while the majority, or 70% of sporadic cancers, have no hereditary component.

In sporadic cancers, a deficiency in DNA repair is occasionally due to a mutation in a DNA repair gene, but much more frequently reduced or absent expression of DNA repair genes is due to epigenetic alterations that reduce or silence gene expression. This is indicated in the figure at the 3rd level from the top. For example, for 113 colorectal cancers examined in sequence, only four had a missense mutation in the DNA repair gene MGMT, while the majority had reduced MGMT expression due to methylation of the MGMT promoter region (an epigenetic alteration). Five reports present evidence that between 40% and 90% of colorectal cancers have reduced MGMT expression due to methylation of the MGMT promoter region.

Similarly, out of 119 cases of mismatch repair-deficient colorectal cancers that lacked DNA repair gene PMS2 expression, PMS2 was deficient in 6 due to mutations in the PMS2 gene, while in 103 cases PMS2 expression was deficient because its pairing partner MLH1 was repressed due to promoter methylation (PMS2 protein is unstable in the absence of MLH1). In the other 10 cases, loss of PMS2 expression was likely due to epigenetic overexpression of the microRNA, miR-155, which down-regulates MLH1.

In further examples tabulated in Cancer epigenetics, epigenetic defects in cancers were found at frequencies of between 13%-100% for the DNA repair genes BRCA1, WRN, FANCB, FANCF, MGMT, MLH1, MSH2, MSH4, ERCC1, XPF, NEIL1 and ATM in cancers including those in breast, ovarian, colorectal, and the head and neck areas. In particular, two or more epigenetic deficiencies in expression of ERCC1, XPF and/or PMS2 were shown to occur simultaneously in the majority of the 49 colon cancers evaluated by Facista et al.

When expression of DNA repair genes is reduced, this causes a DNA repair deficiency. This is shown in the figure at the 4th level from the top. With a DNA repair deficiency, more DNA damages remain in cells at a higher than usual level (5th level from the top in figure), and these excess damages cause increased frequencies of mutation and/or epimutation (6th level from top of figure). Experimentally, mutation rates increase substantially in cells defective in DNA mismatch repair or in Homologous recombinational repair (HRR). Chromosomal rearrangements and aneuploidy also increase in HRR defective cells During repair of DNA double strand breaks, or repair of other DNA damages, incompletely cleared sites of repair can cause epigenetic gene silencing.

Many studies of heavy metal-induced carcinogenesis show that such heavy metals cause reduction in expression of DNA repair enzymes, some through epigenetic mechanisms. In some cases, DNA repair inhibition is proposed to be a predominant mechanism in heavy metal-induced carcinogenicity. For example, one group of studies shows that arsenic inhibits the DNA repair genes PARP, XRCC1, Ligase 3, Ligase 4, DNA POLB, XRCC4, DNA PKCS, TOPO2B, OGG1, ERCC1, XPF, XPB, XPC XPE and P53. Another group of studies shows that cadmium inhibits the DNA repair genes MSH2, ERCC1, XRCC1, OGG1, MSH6, DNA-PK, XPD and XPC

The somatic mutations and epigenetic alterations caused by DNA damages and deficiencies in DNA repair accumulate in field defects. Field defects are normal appearing tissues with multiple alterations (discussed in the section below), and are common precursors to development of the disordered and improperly proliferating clone of tissue in a cancer. Such field defects (second level from bottom of figure) may have multiple mutations and epigenetic alterations.

It is impossible to determine the initial cause for most specific cancers. In a few cases, only one cause exists; for example, the virus HHV-8 causes all Kaposi›s sarcomas. However, with the help of cancer epidemiology techniques and information, it is possible to produce an estimate of a likely cause in many more situations. For example, lung cancer has several causes, including tobacco use and radon gas. Men who currently smoke tobacco develop lung cancer at a rate 14 times that of men who have never smoked tobacco, so the chance of lung cancer in a current smoker being caused by smoking is about 93%; there is a 7% chance that the smoker›s lung cancer was caused by radon gas or some other, non-tobacco cause. These statistical correlations have made it possible for researchers to infer that certain substances or behaviors are carcinogenic. Tobacco smoke causes increased exogenous DNA damage, and these DNA damages are the likely cause of lung cancer due to smoking. Among the more than 5,000 compounds in tobacco smoke, the genotoxic DNA damaging agents that occur both at the highest concentrations and which have the strongest mutagenic effects are acrolein, formaldehyde, acrylonitrile, 1,3-butadiene, acetaldehyde, ethylene oxide and isoprene.

Using molecular biological techniques, it is possible to characterize the mutations, epimutations or chromosomal aberrations within a tumor, and rapid progress is being made in the field of predicting prognosis based on the spectrum of mutations in some cases. For example, up to half of all tumors have a defective p53 gene. This mutation is associated with poor prognosis, since those tumor cells are less likely to go into apoptosis or programmed cell death when damaged by therapy. Telomerase mutations remove additional barriers, extending the number of times a cell can divide. Other mutations enable the tumor to grow new blood vessels to provide more nutrients, or to metastasize, spreading to other parts of the body. However, once a cancer is formed it continues to evolve and to produce sub clones. For example, a renal cancer, sampled in 9 areas, had 40 ubiquitous mutations, 59 mutations shared by some, but not all regions, and 29 private mutations only present in one region.

Contribution of Field Defects

Longitudinally opened freshly resected colon segment showing a cancer and four polyps. Plus a schematic diagram indicating a likely field defect (a region of tissue that precedes and predisposes to the development of cancer) in this colon segment. The diagram indicates sub-clones and sub-sub-clones that were precursors to the tumors.

The term "field cancerization" was first used in 1953 to describe an area or "field" of epithelium that has been preconditioned by (at that time) largely unknown processes so as to predispose it towards development of cancer.Since then, the terms "field cancerization" and "field defect" have been used to describe pre-malignant tissue in which new cancers are likely to arise.

Field defects have been identified in association with cancers and are important in progression to cancer. However, it was pointed out by Rubin that "the vast majority of studies in cancer research has been done on well-defined tumors in vivo, or on discrete neoplastic foci in vitro. Yet there is evidence that more than 80% of the somatic mutations found in mutator phenotype human colorectal tumors occur before the onset of terminal clonal expansion… " Similarly, Vogelstein et al. also indicated that more than half of somatic mutations identified in tumors occurred in a pre-neoplastic phase (in a field defect), during growth of apparently normal cells. It would also be expected that many of the epigenetic alterations present in tumors may have occurred in pre-neoplastic field defects.

In the colon, a field defect probably arises by natural selection of a mutant or epigenetically altered cell among the stem cells at the base of one of the intestinal crypts on the inside surface of the colon. A mutant or epigenetically altered stem cell may replace the other nearby stem cells by natural selection. This may cause a patch of abnormal tissue to arise. The figure in this section includes a photo of a freshly resected and lengthwise-opened segment of the colon showing a colon cancer and four polyps. Below the photo there is a schematic diagram of how a large patch of mutant or epigenetically altered cells may have formed, shown by the large area in yellow in the diagram. Within this first large patch in the diagram (a large clone of cells), a second such mutation or epigenetic alteration may occur so that a given stem cell acquires an advantage compared to other stem cells within the patch, and this altered stem cell may expand clonally forming a secondary patch, or sub-clone, within the original patch. This is indicated in the diagram by four smaller patches of different colors within the large yellow original area. Within these new patches (sub-clones), the process may be repeated multiple times, indicated by the still smaller patches within the four secondary patches (with still different colors in the diagram) which clonally expand, until stem cells arise that generate either small polyps or else a malignant neoplasm (cancer). In the photo, an apparent field defect in this segment of a colon has generated four polyps (labeled with the size of the polyps, 6mm, 5mm, and two of 3mm, and a cancer about 3 cm across in its longest dimension). These neoplasms are also indicated (in the diagram below the photo) by 4 small tan circles (polyps) and a larger red area (cancer). The cancer in the photo occurred in the cecal area of the colon, where the colon joins the small intestine (labeled) and where the appendix occurs (labeled). The fat in the photo is external to the outer wall of the colon. In the segment of colon shown here, the colon was cut open lengthwise to expose the inner surface of the colon and to display the cancer and polyps occurring within the inner epithelial lining of the colon.

If the general process by which sporadic colon cancers arise is the formation of a pre-neoplastic clone that spreads by natural selection, followed by formation of internal sub-clones within the initial clone, and sub-sub-clones inside those, then colon cancers generally should be associated with, and be preceded by, fields of increasing abnormality reflecting the succession of premalignant events. The most extensive region of abnormality (the outermost yellow irregular area in the diagram) would reflect the earliest event in formation of a malignant neoplasm.

In experimental evaluation of specific DNA repair deficiencies in cancers, many specific DNA repair deficiencies were also shown to occur in the field defects surrounding those cancers. The Table, below, gives examples for which the DNA repair deficiency in a cancer was shown to be caused by an epigenetic alteration, and the somewhat lower frequencies with which the same epigenetically caused DNA repair deficiency was found in the surrounding field defect.

Frequency of epigenetic changes in DNA repair genes in sporadic cancers and in adjacent field defects

Cancer	Gene	Frequency in Cancer	Frequency in Field Defect	Reference
Colorectal	MGMT	46%	34%	
Colorectal	MGMT	47%	11%	
Colorectal	MGMT	70%	60%	
Colorectal	MSH2	13%	5%	
Colorectal	ERCC1	100%	40%	
Colorectal	PMS2	88%	50%	
Colorectal	XPF	55%	40%	
Head and Neck	MGMT	54%	38%	
Head and Neck	MLH1	33%	25%	
Head and Neck	MLH1	31%	20%	
Stomach	MGMT	88%	78%	
Stomach	MLH1	73%	20%	
Esophagus	MLH1	77%–100%	23%–79%	

Some of the small polyps in the field defect shown in the photo of the opened colon segment may be relatively benign neoplasms. Of polyps less than 10mm in size, found during colonoscopy and followed with repeat colonoscopies for 3 years, 25% were unchanged in size, 35% regressed or shrank in size while 40% grew in size.

Genome Instability

Cancers are known to exhibit genome instability or a mutator phenotype. The protein-coding DNA within the nucleus is about 1.5% of the total genomic DNA.Within this protein-coding DNA (called the exome), an average cancer of the breast or colon can have about 60 to 70 protein altering mutations, of which about 3 or 4 may be driver mutations, and the remaining ones may be passenger mutations. However, the average number of DNA sequence mutations in the entire genome (including non-protein-coding regions) within a breast cancer tissue sample is about 20,000. In an average melanoma tissue sample (where melanomas have a higher exome mutation frequency) the total number of DNA sequence mutations is about 80,000. These high frequencies of mutations in the total nucleotide sequences within cancers suggest that often an early alteration in the field defect giving rise to a cancer (e.g. yellow area in the diagram in the preceding section) is a deficiency in DNA repair. The large field defects surrounding colon cancers (extending to about 10 cm on each side of a cancer) were shown by Facista et al. to frequently have epigenetic defects in 2 or 3 DNA repair proteins (ERCC1, XPF and/or PMS2) in the entire area of the field defect. When expression of DNA repair genes is reduced, DNA damages accumulate in cells at a higher than normal level, and these excess damages cause increased frequencies of mutation and/or epimutation. Mutation rates strongly increase in cells defective in DNA mismatch repair or in homologous recombinational repair (HRR). A deficiency in DNA repair, itself, can allow DNA damages to accumulate, and error-prone translesion synthesis past some of those damages may give rise to mutations. In addition, faulty repair of these accumulated DNA damages may give rise to epimutations. These new mutations and/or epimutations may provide a proliferative advantage, generating a field defect. Although the mutations/epimutations in DNA repair genes do not, themselves, confer a selective advantage, they may be carried along as passengers in cells when the cell acquires an additional mutation/epimutation that does provide a proliferative advantage.

Non-Mainstream Theories

There are a number of theories of carcinogenesis and cancer treatment that fall outside the mainstream of scientific opinion, due to lack of scientific rationale, logic, or evidence base. These theories may be used to justify various alternative cancer treatments. They should be distinguished from those theories of carcinogenesis that have a logical basis within mainstream cancer biology, and from which conventionally testable hypotheses can be made.

Several alternative theories of carcinogenesis, however, are based on scientific evidence and are increasingly being acknowledged. Some researchers believe that cancer may be caused by aneuploidy (numerical and structural abnormalities in chromosomes) rather than by mutations or epimutations. Cancer has also been considered as a metabolic disease in which the cellular metabolism of oxygen is diverted from the pathway that generates energy (oxidative phosphorylation) to the pathway that generates reactive oxygen species (figure). This causes an energy switch from oxidative phosphorylation to aerobic glycolysis (Warburg›s hypothesis) and the accumulation of reactive oxygen species leading to oxidative stress (oxidative stress theory of cancer). All these theories of carcinogenesis may be complementary rather than contradictory. Aberrant DNA methylation patterns – hypermethylation and hypomethylation compared to normal tissue – have been associated with a large number of human malignancies. (See DNA methylation in cancer)

A number of authors have questioned the assumption that cancers result from sequential random mutations as oversimplistic, suggesting instead that cancer results from a failure of the body to inhibit an innate, programmed proliferative tendency.A related theory developed by astrobiologists suggests that cancer is an atavism, an evolutionary throwback to an earlier form of multicellular life. The genes responsible for uncontrolled cell growth and cooperation between cancer cells are very similar to those that enabled the first multicellular life forms to group together and flourish. These genes still exist within the genome of more complex metazoans, such as humans, although more recently evolved genes keep them in check. When the newer controlling genes fail for whatever reason, the cell can revert to its more primitive programming and reproduce out of control.

The theory is an alternative to the notion that cancers begin with rogue cells that undergo evolution within the body. Instead they possess a fixed number of primitive genes that are progressively activated, giving them finite variability.

Cancer cell Biology

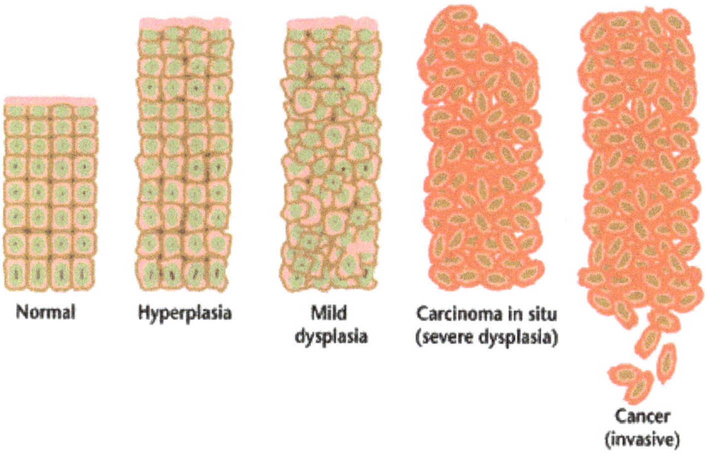

Tissue can be organized in a continuous spectrum from normal to cancer.

Often, the multiple genetic changes that result in cancer may take many years to accumulate. During this time, the biological behavior of the pre-malignant cells slowly change from the properties of normal cells to cancer-like properties. Pre-malignant tissue can have a distinctive appearance under the microscope. Among the distinguishing traits are an increased number of dividing cells, variation in nuclear size and shape, variation in cell size and shape, loss of specialized cell features, and loss of normal tissue organization. Dysplasia is an abnormal type of excessive cell proliferation characterized by loss of normal tissue arrangement and cell structure in pre-malignant cells. These early neoplastic changes must be distinguished from hyperplasia, a reversible increase in cell division caused by an external stimulus, such as a hormonal imbalance or chronic irritation.

The most severe cases of dysplasia are referred to as carcinoma in situ. In Latin, the term in situ means in place, so carcinoma in situ refers to an uncontrolled growth of cells that remains in the original location and has not shown invasion into other tissues. Nevertheless, carcinoma in situ may develop into an invasive malignancy and is usually removed surgically, if possible.

Clonal Evolution

Just like a population of animals undergoes evolution, an unchecked population of cells also can undergo evolution. This undesirable process is called somatic evolution, and is how cancer arises and becomes more malignant.

Most changes in cellular metabolism that allow cells to grow in a disorderly fashion lead to cell death. However once cancer begins, cancer cells undergo a process of natural selection: the few cells with new genetic changes that enhance their survival or reproduction continue to multiply, and soon come to dominate the growing tumor, as cells with less favorable genetic change are out-competed. This is exactly how pathogens such as MRSA can become antibiotic-resistant (or how HIV can become drug-resistant), and the same reason why crop blights and pests can become pesticide-resistant. This evolution is why cancer recurrences will have cells that have acquired cancer-drug resistance (or in some cases, resistance to radiation from radiotherapy).

Biological Properties of Cancer Cells

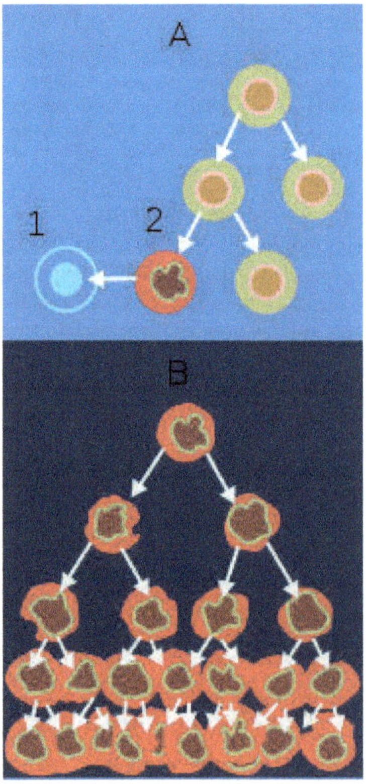

When normal cells are damaged beyond repair, they are eliminated by apoptosis (A). Cancer cells avoid apoptosis and continue to multiply in an unregulated manner (B).

In a 2000 article by Hanahan and Weinberg, the biological properties of malignant tumor cells were summarized as follows:

- Acquisition of self-sufficiency in growth signals, leading to unchecked growth.
- Loss of sensitivity to anti-growth signals, also leading to unchecked growth.
- Loss of capacity for apoptosis, in order to allow growth despite genetic errors and external anti-growth signals.

- Loss of capacity for senescence, leading to limitless replicative potential (immortality)
- Acquisition of sustained angiogenesis, allowing the tumor to grow beyond the limitations of passive nutrient diffusion.
- Acquisition of ability to invade neighbouring tissues, the defining property of invasive carcinoma.
- Acquisition of ability to build metastases at distant sites, the classical property of malignant tumors (carcinomas or others).

The completion of these multiple steps would be a very rare event without :

- Loss of capacity to repair genetic errors, leading to an increased mutation rate (genomic instability), thus accelerating all the other changes.

These biological changes are classical in carcinomas; other malignant tumors may not need to achieve them all. For example, tissue invasion and displacement to distant sites are normal properties of leukocytes; these steps are not needed in the development of leukemia. The different steps do not necessarily represent individual mutations. For example, inactivation of a single gene, coding for the p53 protein, will cause genomic instability, evasion of apoptosis and increased angiogenesis. Not all the cancer cells are dividing. Rather, a subset of the cells in a tumor, called cancer stem cells, replicate themselves and generate differentiated cells.

Cancer as a Defect in Cell Interactions

Normally, once a tissue is injured or infected, damaged cells elicit inflammation, by stimulating specific patterns of enzyme activity and cytokine gene expression on surrounding cells. Discrete clusters of molecules are secreted, which act as mediators, inducing the activity of subsequent cascades of biochemical changes.Each cytokine binds to specific receptors on various cell types, and each cell type responds differently by altering the activity of intracellular signal transduction pathways, depending on the receptors that the cell expresses and the signaling molecules present inside the cell. Collectively, this reprogramming process induces a stepwise change in cell phenotypes, which will ultimately lead to restoration of tissue function and toward regaining essential structural integrity.A tissue can thereby heal, depending on the productive communication between the cells present at the site of damage, and the immune system. Key factor in healing is the regulation of cytokine gene expression, which enables complementary groups of cells to respond to inflammatory mediators in a manner that gradually produces essential changes in tissue physiology. Cancer cells have either permanent (genetic) or reversible (epigenetic) changes on their genome, which partly inhibit their communication with surrounding cells and with the immune system. Cancer cells do not communicate with their tissue microenvironment in a manner that protects tissue integrity; instead, the movement and the survival of cancer cells become possible in locations where they can impair tissue function. Cancer cells survive by rewiring signal pathways that normally protect the tissue from the immune system.

One example for rewiring of tissue function in cancer is the activity of transcription factor NF-κB. NF-κB activates the expression of numerous genes that are involved in the transition between inflammation and regeneration, which encode cytokines, adhesion factors, and other molecules that can change cell fate. This reprogramming of cellular phenotypes normally allows the development of a fully functional intact tissue. NF-κB activity is tightly controlled by multiple proteins, which collectively ensure that only discrete clusters of genes are induced by NF-κB in a given cell and at a given time. This tight regulation of signal exchange between cells, protects the tissue from excessive inflammation, and ensures that different cell types would gradually acquire complementary functions, and specific positions. Failure of this mutual regulation between genetic reprogramming and cell interactions allows cancer cells to give rise to metastasis. Cancer cells respond aberrantly to cytokines, and activate signal cascades that can protect them from the immune system.

In Fishes

The role of iodine in marine fishes (rich in iodine) and freshwater fishes (iodine-deficient) is not completely understood, but it has been reported that freshwater fishes are more susceptible to infectious and, in particular, neoplastic and atherosclerotic diseases, of marine fishes. Walsh, Luer et al. of the Marine Biomedical Research Program of the Mote Marine Laboratory in Sarasota, and in 1999, the Senate Committee concerning the "Marine Research" have reported that marine elasmobranch fishes such as sharks, stingrays etc. are much less affected by cancer than freshwater fishes, and therefore have stimulated medical research to better understand carcinogenesis so it can be useful in other animals and especially in humans.

Mechanisms

In order for cells to start dividing uncontrollably, genes that regulate cell growth must be dysregulated. Proto-oncogenes are genes that promote cell growth and mitosis, whereas tumor suppressor genes discourage cell growth, or temporarily halt cell division to carry out DNA repair. Typically, a series of several mutations to these genes is required before a normal cell transforms into a cancer cell. This concept is sometimes termed "oncoevolution." Mutations to these genes provide the signals for tumor cells to start dividing uncontrollably. But the uncontrolled cell division that characterizes cancer also requires that the dividing cell duplicates all its cellular components to create two daughter cells. The activation of anaerobic glycolysis (the Warburg effect), which is not necessarily induced by mutations in proto-oncogenes and tumor suppressor genes, provides most of the building blocks required to duplicate the cellular components of a dividing cell and, therefore, is also essential for carcinogenesis.

Cell Types Involved in Cancer Growth

There are several different cell types that are critical to tumour growth. In particular endothelial progenitor cells are a very important cell population in tumour blood vessel growth. The hypothesis that endothelial progenitor cells are important in tumour growth, angiogenesis and metastasis has been supported by a recent publication in Cancer Research (August 2010). This paper argues that endothelial progenitor cells can be marked using the Inhibitor of DNA Binding 1 (ID1). This novel finding meant that investigators were able to track endothelial progenitor cells from the bone marrow to the blood to the tumour-stroma and vasculature. This finding of endothelial progenitor cells incorporated in tumour vasculature gives evidence for the importance of this cell type in blood vessel development in a tumour setting and metastasis. Furthermore, ablation of the endothelial progenitor cells in the bone marrow lead to a significant decrease in tumour growth and vasculature development. The continued research into the importance of endothelial progenitor cells may present novel therapeutic targets.

Oncogenes

Oncogenes promote cell growth through a variety of ways. Many can produce hormones, a chemical messenger between cells that encourage mitosis, the effect of which depends on the signal transduction of the receiving tissue or cells. In other words, when a hormone receptor on a recipient cell is stimulated, the signal is conducted from the surface of the cell to the cell nucleus to affect some change in gene transcription regulation at the nuclear level. Some oncogenes are part of the signal transduction system itself, or the signal receptors in cells and tissues themselves, thus controlling the sensitivity to such hormones. Oncogenes often produce mitogens, or are involved in transcription of DNA in protein synthesis, which creates the proteins and enzymes responsible for producing the products and biochemicals cells use and interact with.

Mutations in proto-oncogenes, which are the normally quiescent counterparts of oncogenes, can modify their expression and function, increasing the amount or activity of the product protein. When this happens, the proto-oncogenes become oncogenes, and this transition upsets the normal balance of cell cycle regulation in the cell, making uncontrolled growth possible. The chance of cancer cannot be reduced by removing proto-oncogenes from the genome, even if this were possible, as they are critical for growth, repair and homeostasis of the organism. It is only when they become mutated that the signals for growth become excessive.

One of the first oncogenes to be defined in cancer research is the ras oncogene. Mutations in the Ras family of proto-oncogenes (comprising H-Ras, N-Ras and K-Ras) are very common, being found in 20% to 30% of all human tumours. Ras was originally identified in the Harvey sarcoma virus genome, and researchers were surprised that not only is this gene present in the human genome but also, when ligated to a stimulating control element, it could induce cancers in cell line cultures.

Proto-Oncogenes

Proto-oncogenes promote cell growth in a variety of ways. Many can produce hormones, "chemical messengers" between cells that encourage mitosis, the effect of which depends on the signal transduction of the receiving tissue or cells. Some are responsible for the signal transduction system and signal receptors in cells and tissues themselves, thus controlling the sensitivity to such hormones. They often produce mitogens, or are involved in transcription of DNA in protein synthesis, which create the proteins and enzymes is responsible for producing the products and biochemicals cells use and interact with.

Mutations in proto-oncogenes can modify their expression and function, increasing the amount or activity of the product protein. When this happens, they become oncogenes, and, thus, cells have a higher chance to divide excessively and uncontrollably. The chance of cancer cannot be reduced by removing proto-oncogenes from the genome, as they are critical for growth, repair and homeostasis of the body. It is only when they become mutated that the signals for growth become excessive. It is important to note that a gene possessing a growth-promoting role may increase carcinogenic potential of a cell, under the condition that all necessary cellular mechanisms that permit growth are activated. This condition includes also the inactivation of specific tumor suppressor genes (see below). If the condition is not fulfilled, the cell may cease to grow and can proceed to die. This makes knowledge of the stage and type of cancer cell that grows under the control of a given oncogene crucial for the development of treatment strategies.

Tumor Suppressor Genes

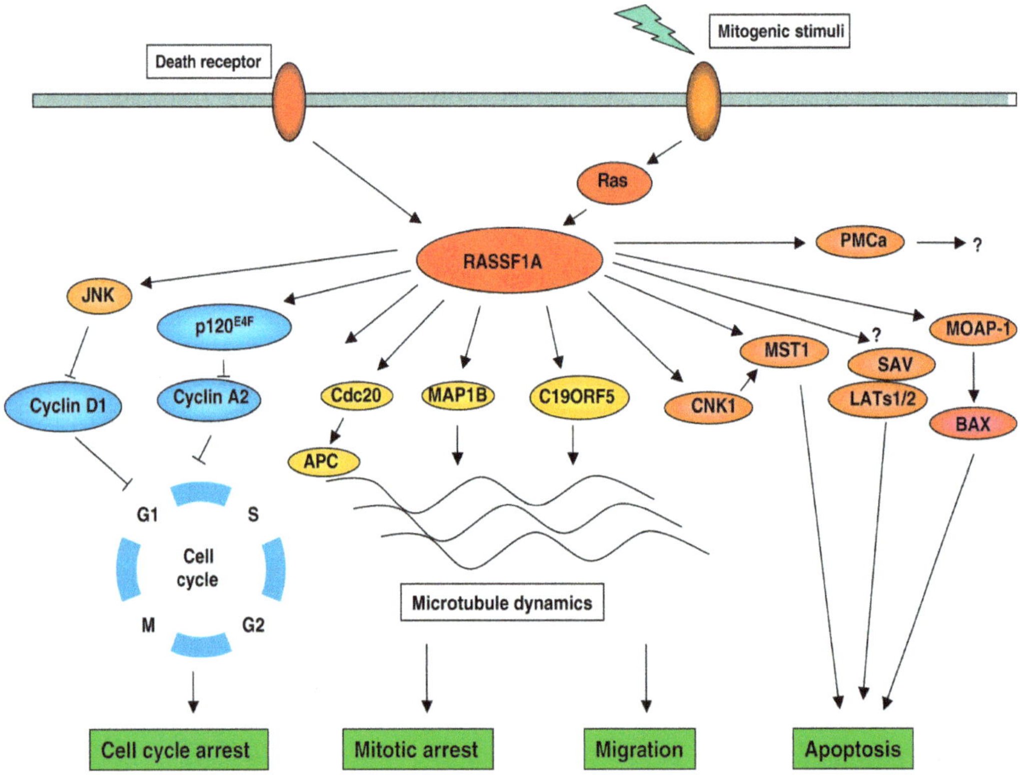

Many tumor suppressor genes effect signal transduction pathways that regulate apoptosis, also known as "programmed cell death".

Tumor suppressor genes code for anti-proliferation signals and proteins that suppress mitosis and cell growth. Generally, tumor suppressors are transcription factors that are activated by cellular stress or DNA damage. Often DNA damage will cause the presence of free-floating genetic material as well as other signs, and will trigger enzymes and pathways that lead to the activation of tumor suppressor genes. The functions of such genes is to arrest the progression of the cell cycle in order to carry out DNA repair, preventing mutations from being passed on to daughter cells. The p53 protein, one of the most important studied tumor suppressor genes, is a transcription factor activated by many cellular stressors including hypoxia and ultraviolet radiation damage.

Despite nearly half of all cancers possibly involving alterations in p53, its tumor suppressor function is poorly understood. p53 clearly has two functions: one a nuclear role as a transcription factor, and the other a cytoplasmic role in regulating the cell cycle, cell division, and apoptosis.

The Warburg hypothesis is the preferential use of glycolysis for energy to sustain cancer growth. p53 has been shown to regulate the shift from the respiratory to the glycolytic pathway.

However, a mutation can damage the tumor suppressor gene itself, or the signal pathway that activates it, switching it off. The invariable consequence of this is that DNA repair is hindered or inhibited: DNA damage accumulates without repair, inevitably leading to cancer.

Mutations of tumor suppressor genes that occur in germline cells are passed along to offspring, and increase the likelihood for cancer diagnoses in subsequent generations. Members of these families have increased incidence and decreased latency of multiple tumors. The tumor types are typical for each type of tumor suppressor gene mutation, with some mutations causing particular cancers, and other mutations causing others. The mode of inheritance of mutant tumor suppressors is that an affected member inherits a defective copy from one parent, and a normal copy from the other. For instance, individuals who inherit one mutant *p53* allele (and are therefore heterozygous for mutated *p53*) can develop melanomas and pancreatic cancer, known as Li-Fraumeni syndrome. Other inherited tumor suppressor gene syndromes include *Rb* mutations, linked to retinoblastoma, and *APC* gene mutations, linked to adenopolyposis colon cancer. Adenopolyposis colon cancer is associated with thousands of polyps in colon while young, leading to colon cancer at a relatively early age. Finally, inherited mutations in *BRCA1* and *BRCA2* lead to early onset of breast cancer.

Development of cancer was proposed in 1971 to depend on at least two mutational events. In what became known as the Knudson two-hit hypothesis, an inherited, germ-line mutation in a tumor suppressor gene would cause cancer only if another mutation event occurred later in the organism's life, inactivating the other allele of that tumor suppressor gene.

Usually, oncogenes are dominant, as they contain gain-of-function mutations, while mutated tumor suppressors are recessive, as they contain loss-of-function mutations. Each cell has two copies of the same gene, one from each parent, and under most cases gain of function mutations in just one copy of a particular proto-oncogene is enough to make that gene a true oncogene. On the other hand, loss of function mutations need to happen in both copies of a tumor suppressor gene to render that gene completely non-functional. However, cases exist in which one mutated copy of a tumor suppressor gene can render the other, wild-type copy non-functional. This phenomenon is called the *dominant negative effect* and is observed in many p53 mutations.

Knudson›s two hit model has recently been challenged by several investigators. Inactivation of one allele of some tumor suppressor genes is sufficient to cause tumors. This phenomenon is called haploinsufficiency and has been demonstrated by a number of experimental approaches. Tumors caused by haploinsufficiency usually have a later age of onset when compared with those by a two hit process.

Multiple Mutations

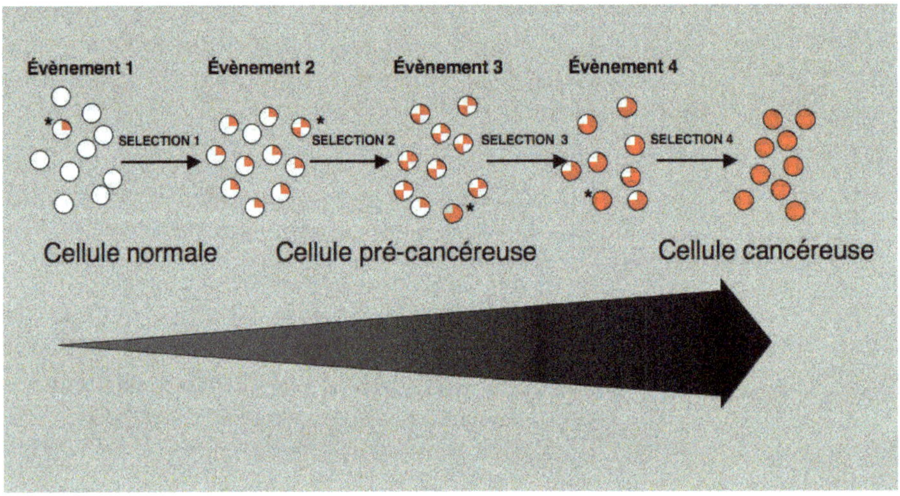

Multiple mutations in cancer cells.

In general, mutations in both types of genes are required for cancer to occur. For example, a mutation limited to one oncogene would be suppressed by normal mitosis control and tumor suppressor genes, first hypothesised by the Knudson hypothesis.A mutation to only one tumor suppressor gene would not cause cancer either, due to the presence of many backup genes that duplicate its functions. It is only when enough proto-oncogenes have mutated into oncogenes, and enough tumor suppressor genes deactivated or damaged, that the signals for cell growth overwhelm the signals to regulate it, that cell growth quickly spirals out of control. Often, because these genes regulate the processes that prevent most damage to genes themselves, the rate of mutations increases as one gets older, because DNA damage forms a feedback loop.

Mutation of tumor suppressor genes that are passed on to the next generation of not merely cells, but their offspring, can cause increased likelihoods for cancers to be inherited. Members within these families have increased incidence and decreased latency of multiple tumors. The mode of inheritance of mutant tumor suppressors is that affected member inherits a defective copy from one parent, and a normal copy from another. Because mutations in tumor suppressors act in a recessive manner (note, however, there are exceptions), the loss of the normal copy creates the cancer phenotype. For instance, individuals that are heterozygous for p53 mutations are often victims of Li-Fraumeni syndrome, and that are heterozygous for Rb mutations develop retinoblastoma. In similar fashion, mutations in the adenomatous polyposis coli gene are linked to adenopolyposis colon cancer, with thousands of polyps in the colon while young, whereas mutations in BRCA1 and BRCA2 lead to early onset of breast cancer.

A new idea announced in 2011 is an extreme version of multiple mutations, called *chromothripsis* by its proponents. This idea, affecting only 2–3% of cases of cancer, although up to 25% of bone cancers, involves the catastrophic shattering of a chromosome into tens or hundreds of pieces and then being patched back together incorrectly. This shattering probably takes place when the chromosomes are compacted during normal cell division, but the trigger for the shattering is unknown. Under this model, cancer arises as the result of a single, isolated event, rather than the slow accumulation of multiple mutations.

Non-Mutagenic Carcinogens

Many mutagens are also carcinogens, but some carcinogens are not mutagens. Examples of carcinogens that are not mutagens include alcohol and estrogen. These are thought to promote cancers through their stimulating effect on the rate of cell mitosis. Faster rates of mitosis increasingly leave fewer opportunities for repair enzymes to repair damaged DNA during DNA replication, increasing the likelihood of a genetic mistake. A mistake made during mitosis can lead to the daughter cells› receiving the wrong number of chromosomes, which leads to aneuploidy and may lead to cancer.

Role of infections

Bacterial

Viral

Furthermore, many cancers originate from a viral infection; this is especially true in animals such as birds, but less so in humans. 12% of human cancers can be attributed to a viral infection. The mode of virally induced tumors can be divided into two, *acutely transforming* or *slowly transforming*. In acutely transforming viruses, the viral particles carry a gene that encodes for an overactive oncogene called viral-oncogene (v-onc), and the infected cell is transformed as soon as v-onc is expressed. In contrast, in slowly transforming viruses, the virus genome is inserted, especially as viral genome insertion is obligatory part of retroviruses, near a proto-oncogene in the host genome. The viral promoter or other transcription regulation elements, in turn, cause over-expression of that proto-oncogene, which, in turn, induces uncontrolled cellular proliferation. Because viral genome insertion is not specific to proto-oncogenes and the chance of insertion near that proto-oncogene is low, slowly transforming viruses have very long tumor latency compared to acutely transforming virus, which already carries the viral-oncogene.

Viruses that are known to cause cancer such as HPV (cervical cancer), Hepatitis B (liver cancer), and EBV (a type of lymphoma), are all DNA viruses. It is thought that when the virus infects a cell, it inserts a part of its own DNA near the cell growth genes, causing cell division. The group of changed cells that are formed from the first cell dividing all have the same viral DNA near the cell growth genes. The group of changed cells are now special because one of the normal controls on growth has been lost.

Depending on their location, cells can be damaged through radiation from sunshine, chemicals from cigarette smoke, and inflammation from bacterial infection or other viruses. Each cell has a chance of damage, a step on a path toward cancer. Cells often die if they are damaged, through failure of a vital process or the immune system; however, sometimes damage will knock out a single cancer gene. In an old person, there are thousands, tens of thousands or hundreds of thousands of knocked-out cells. The chance that any one would form a cancer is very low.

When the damage occurs in any area of changed cells, something different occurs. Each of the cells has the potential for growth. The changed cells will divide quicker when the area is damaged by physical, chemical, or viral agents. A vicious circle has been set up: Damaging the area will cause the changed cells to divide, causing a greater likelihood that they will suffer knock-outs.

This model of carcinogenesis is popular because it explains why cancers grow. It would be expected that cells that are damaged through radiation would die or at least be worse off because they have fewer genes working; viruses increase the number of genes working.

One concern is that we may end up with thousands of vaccines to prevent every virus that can change our cells. Viruses can have different effects on different parts of the body. It may be possible to prevent a number of different cancers by immunizing against one viral agent. It is likely that HPV, for instance, has a role in cancers of the mucous membranes of the mouth.

Helminthiasis

Certain parasitic worms are known to be carcinogenic. These include:

- *Clonorchis sinensis* (the organism causing Clonorchiasis) and *Opisthorchis viverrini* (causing Opisthorchiasis) are associated with cholangiocarcinoma.
- *Schistosoma species* (the organisms causing Schistosomiasis) is associated with bladder cancer.

Epigenetics

Epigenetics is the study of the regulation of gene expression through chemical, non-mutational changes in DNA structure. The theory of epigenetics in cancer pathogenesis is that non-mutational changes to DNA can lead to alterations in gene expression. Normally, oncogenes are silent, for example, because of DNA methylation. Loss of that methylation can induce the aberrant expression of oncogenes, leading to cancer pathogenesis. Known mechanisms of epigenetic change include DNA methylation, and methylation or acetylation of histone proteins bound to chromosomal DNA at specific locations. Classes of medications, known as HDAC inhibitors and DNA methyltransferase inhibitors, can re-regulate the epigenetic signaling in the cancer cell.

Cancer Stem Cells

A new way of looking at carcinogenesis comes from integrating the ideas of developmental biology into oncology. The cancer stem cell hypothesis proposes that the different kinds of cells in a heterogeneous tumor arise from a single cell, termed Cancer Stem Cell. Cancer stem cells may arise from transformation of adult stem cells or differentiated cells within a body. These cells persist as a subcomponent of the tumor and retain key stem cell properties. They give rise to a variety of cells, are capable of self-renewal and homeostatic control. Furthermore, the relapse of cancer and the emergence of metastasis are also attributed to these cells. The cancer stem cell hypothesis does not contradict earlier concepts of carcinogenesis. The cancer stem cell hypothesis has been a proposed mechanism that contributes to tumour heterogeneity.

Clonal Evolution

While genetic and epigenetic alterations in tumor suppressor genes and oncogenes change the behavior of cells, those alterations, in the end, result in cancer through their effects on the population of neoplastic cells and their microenvironment.Mutant cells in neoplasms compete for space and resources. Thus, a clone with a mutation in a tumor suppressor gene or oncogene will expand only in a neoplasm if that mutation gives the clone a competitive advantage over the other clones and normal cells in its microenvironment. Thus, the process of carcinogenesis is formally a process of Darwinian evolution, known as somatic or clonal evolution.Furthermore, in light of the Darwinistic mechanisms of carcinogenesis, it has been theorized that the various forms of cancer can be categorized as pubertarial and gerontological. Anthropological research is currently being conducted on cancer as a natural evolutionary process through which natural selection destroys environmentally inferior phenotypes while supporting others. According to this theory, cancer comes in two separate types: from birth to the end of puberty (approximately age 20) teleologically inclined toward supportive group dynamics, and from mid-life to death (approximately age 40+) teleologically inclined away from overpopulative group dynamics.

Mechanical Factors

There is such point of view that mechanical factors can cause carcinogenesis.

REFERENCES

1. Fearon ER, Vogelstein B (June 1990). "A genetic model for colorectal tumorigenesis". Cell. 61 (5): 759–67. doi:10.1016/0092-8674(90)90186-I. PMID 2188735.
2. Wood LD, Parsons DW, Jones S, Lin J, Sjöblom T, Leary RJ, Shen D, Boca SM, Barber T, Ptak J, Silliman N, Szabo S, Dezso Z, Ustyanksky V, Nikolskaya T, Nikolsky Y, Karchin R, Wilson PA, Kaminker JS, Zhang Z, Croshaw R, Willis J, Dawson D, Shipitsin M, Willson JK, Sukumar S, Polyak K, Park BH, Pethiyagoda CL, Pant PV, Ballinger DG, Sparks AB, Hartigan J, Smith DR,

Suh E, Papadopoulos N, Buckhaults P, Markowitz SD, Parmigiani G, Kinzler KW, Velculescu VE, Vogelstein B (2007). "The genomic landscapes of human breast and colorectal cancers". Science. 318 (5853): 1108–1113. doi:10.1126/science.1145720. PMID 17932254.

3. Croce CM (January 2008). "Oncogenes and cancer". The New England Journal of Medicine. 358 (5): 502–11. doi:10.1056/NEJMra072367. PMID 18234754.
4. Lim, LP; Lau, NC; Garrett-Engele, P; Grimson, A; Schelter, JM; Castle, J; Bartel, DP; Linsley, PS; Johnson, JM (2005). "Microarray analysis shows that some microRNAs downregulate large numbers of target mRNAs". Nature. 433 (7027): 769–773. doi:10.1038/nature03315. PMID 15685193.
5. Balaguer, F.; Link, A.; Lozano, J. J.; Cuatrecasas, M.; Nagasaka, T.; Boland, C. R.; Goel, A. (2010). "Epigenetic Silencing of miR-137 is an Early Event in Colorectal Carcinogenesis". Cancer Research. 70 (16): 6609–6618. doi:10.1158/0008-5472.CAN-10-0622. PMC 2922409 . PMID 20682795.
6. Knudson AG (November 2001). "Two genetic hits (more or less) to cancer". Nature Reviews Cancer. 1 (2): 157–62. doi:10.1038/35101031. PMID 11905807.
7. Daniel, Filipe Ivan; Cherubini, Karen; Yurgel, Liliane Soares; De Figueiredo, Maria Antonia Zancanaro; Salum, Fernanda Gonçalves (2011). "The role of epigenetic transcription repression and DNA methyltransferases in cancer". Cancer. 117 (4): 677–687. doi:10.1002/cncr.25482. PMID 20945317.Review.
8. Kanwal, R; Gupta, S (2012). "Epigenetic modifications in cancer". Clinical Genetics. 81 (4): 303–311. doi:10.1111/j.1399-0004.2011.01809.x. PMC 3590802 . PMID 22082348.
9. Pattani, Kavita M.; Soudry, Ethan; Glazer, Chad A.; Ochs, Michael F.; Wang, Hao; Schussel, Juliana; Sun, Wenyue; Hennessey, Patrick; Mydlarz, Wojciech; Loyo, Myriam; Demokan, Semra; Smith, Ian M.; Califano, Joseph A. (2012). Tao, Qian, ed. "MAGEB2 is Activated by Promoter Demethylation in Head and Neck Squamous Cell Carcinoma". PLoS ONE. 7 (9): e45534. doi:10.1371/journal.pone.0045534. PMC 3454438 . PMID 23029077.
10. Sampath, D.; Liu, C.; Vasan, K.; Sulda, M.; Puduvalli, V. K.; Wierda, W. G.; Keating, M. J. (2011). "Histone deacetylases mediate the silencing of miR-15a, miR-16, and miR-29b in chronic lymphocytic leukemia". Blood. 119 (5): 1162–1172. doi:10.1182/blood-2011-05-351510. PMC 3277352 . PMID 22096249.
11. Hitchler, Michael J.; Oberley, Larry W.; Domann, Frederick E. (2008). "Epigenetic silencing of SOD2 by histone modifications in human breast cancer cells". Free Radical Biology and Medicine. 45 (11): 1573–1580. doi:10.1016/j.freeradbiomed.2008.09.005. PMC 2633123 . PMID 18845242.
12. Baldassarre, G; Battista, S; Belletti, B; Thakur, S; Pentimalli, F; Trapasso, F; Fedele, M; Pierantoni, G; Croce, CM; Fusco, A (2003). "Negative regulation of BRCA1 gene expression by HMGA1 proteins accounts for the reduced BRCA1 protein levels in sporadic breast carcinoma". Molecular and Cellular Biology. 23 (7): 2225–2238. doi:10.1128/MCB.23.7.2225-2238.2003. PMC 150734 . PMID 12640109.

13. Lopez-Lazaro M (August 2015). "Stem cell division theory of cancer". Cell Cycle. 14 (16): 1–2. doi:10.1080/15384101.2015.1062330. PMID 26090957.

14. Lopez-Lazaro M (May 2015). "The migration ability of stem cells can explain the existence of cancer of unknown primary site. Rethinking metastasis.". Oncoscience. 2 (5): 467–75. doi:10.18632/oncoscience.159. PMC 4468332 . PMID 26097879.

15. Tomasetti C, Vogelstein B (January 2015). "Cancer etiology. Variation in cancer risk among tissues can be explained by the number of stem cell divisions.". Science. 347 (6217): 78–81. doi:10.1126/science.1260825. PMID 25554788.

16. Bernstein C, Prasad AR, Nfonsam V, Bernstein H. (2013). DNA Damage, DNA Repair and Cancer, New Research Directions in DNA Repair, Prof. Clark Chen (Ed.), ISBN 978-953-51-1114-6, InTech, http://www.intechopen.com/books/new-research-directions-in-dna-repair/dna-damage-dna-repair-and-cancer

17. Kastan MB (2008). "DNA damage responses: mechanisms and roles in human disease: 2007 G.H.A. Clowes Memorial Award Lecture". Mol. Cancer Res. 6 (4): 517–24. doi:10.1158/1541-7786.MCR-08-0020. PMID 18403632.

18. Cunningham, F.H.; Fiebelkorn, S.; Johnson, M.; Meredith, C. (2011). "A novel application of the Margin of Exposure approach: Segregation of tobacco smoke toxicants". Food and Chemical Toxicology. 49 (11): 2921–2933. doi:10.1016/j.fct.2011.07.019. PMID 21802474.

19. Kanavy, Holly E.; Gerstenblith, Meg R. (2011). "Ultraviolet Radiation and Melanoma". Seminars in Cutaneous Medicine and Surgery. 30 (4): 222–228. doi:10.1016/j.sder.2011.08.003. PMID 22123420.

20. Handa, Osamu; Naito, Yuji; Yoshikawa, Toshikazu (2011). "Redox biology and gastric carcinogenesis: The role of Helicobacter pylori". Redox Report. 16 (1): 1–7. doi:10.1179/174329211X12968219310756. PMID 21605492.

21. Smela, ME; Hamm, ML; Henderson, PT; Harris, CM; Harris, TM; Essigmann, JM (2002). "The aflatoxin B(1) formamidopyrimidine adduct plays a major role in causing the types of mutations observed in human hepatocellular carcinoma". Proceedings of the National Academy of Sciences of the United States of America. 99 (10): 6655–60. doi:10.1073/pnas.102167699. PMC 124458 . PMID 12011430.

22. Katsurano, M; Niwa, T; Yasui, Y; Shigematsu, Y; Yamashita, S; Takeshima, H; Lee, M S; Kim, Y-J; Tanaka, T; Ushijima, T (2011). "Early-stage formation of an epigenetic field defect in a mouse colitis model, and non-essential roles of T- and B-cells in DNA methylation induction". Oncogene. 31 (3): 342–351. doi:10.1038/onc.2011.241. PMID 21685942.

23. Bernstein, Carol; Holubec, Hana; Bhattacharyya, Achyut K.; Nguyen, Huy; Payne, Claire M.; Zaitlin, Beryl; Bernstein, Harris (2011). "Carcinogenicity of deoxycholate, a secondary bile acid". Archives of Toxicology. 85 (8): 863–71. doi:10.1007/s00204-011-0648-7. PMC 3149672 . PMID 21267546.

24. Malkin, D. (2011). "Li-Fraumeni Syndrome". Genes & Cancer. 2 (4): 475–484. doi:10.1177/1947601911413466. PMC 3135649 . PMID 21779515.

25. Fearon, ER (1997). "Human cancer syndromes: Clues to the origin and nature

of cancer". Science. 278 (5340): 1043–1050. doi:10.1126/science.278.5340.1043. PMID 9353177.

26. Lichtenstein, Paul; Holm, Niels V.; Verkasalo, Pia K.; Iliadou, Anastasia; Kaprio, Jaakko; Koskenvuo, Markku; Pukkala, Eero; Skytthe, Axel; Hemminki, Kari (2000). "Environmental and Heritable Factors in the Causation of Cancer — Analyses of Cohorts of Twins from Sweden, Denmark, and Finland". New England Journal of Medicine. 343 (2): 78–85. doi:10.1056/NEJM200007133430201. PMID 10891514.
27. Halford, S; Rowan, A; Sawyer, E; Talbot, I; Tomlinson, I (2005). "O6-methylguanine methyltransferase in colorectal cancers: Detection of mutations, loss of expression, and weak association with G:C>A:T transitions". Gut. 54 (6): 797–802. doi:10.1136/gut.2004.059535. PMC 1774551 . PMID 15888787.
28. Shen, L.; Kondo, Y.; Rosner, G. L.; Xiao, L.; Hernandez, N. S.; Vilaythong, J.; Houlihan, P. S.; Krouse, R. S.; Prasad, A. R.; Einspahr, J. G.; Buckmeier, J.; Alberts, D. S.; Hamilton, S. R.; Issa, J.-P. J. (2005). "MGMT Promoter Methylation and Field Defect in Sporadic Colorectal Cancer". JNCI Journal of the National Cancer Institute. 97 (18): 1330–1338. doi:10.1093/jnci/dji275. PMID 16174854.
29. Psofaki, Vasiliki; Kalogera, C; Tzambouras, N; Stephanou, D; Tsianos, E; Seferiadis, K; Kolios, G (2010). "Promoter methylation status ofhMLH1,MGMT, andCDKN2A/p16in colorectal adenomas". World Journal of Gastroenterology. 16 (28): 3553–3560. doi:10.3748/wjg.v16.i28.3553. PMC 2909555 . PMID 20653064.
30. Lee, Kyung-Hwa; Lee, Ji-Shin; Nam, Jong-Hee; Choi, Chan; Lee, Min-Cheol; Park, Chang-Soo; Juhng, Sang-Woo; Lee, Jae-Hyuk (2011). "Promoter methylation status of hMLH1, hMSH2, and MGMT genes in colorectal cancer associated with adenoma–carcinoma sequence". Langenbeck's Archives of Surgery. 396 (7): 1017–1026. doi:10.1007/s00423-011-0812-9. PMID 21706233.
31. Amatu, A.; Sartore-Bianchi, A.; Moutinho, C.; Belotti, A.; Bencardino, K.; Chirico, G.; Cassingena, A.; Rusconi, F.; Esposito, A.; Nichelatti, M.; Esteller, M.; Siena, S. (2013). "Promoter CpG Island Hypermethylation of the DNA Repair Enzyme MGMT Predicts Clinical Response to Dacarbazine in a Phase II Study for Metastatic Colorectal Cancer". Clinical Cancer Research. 19 (8): 2265–72. doi:10.1158/1078-0432.CCR-12-3518. PMID 23422094.
32. Mokarram, P.; Zamani, M.; Kavousipour, S.; Naghibalhossaini, F.; Irajie, C.; Moradi Sarabi, M.; Hosseini, S. V. (2012). "Different patterns of DNA methylation of the two distinct O6-methylguanine-DNA methyltransferase (O6-MGMT) promoter regions in colorectal cancer". Molecular Biology Reports. 40 (5): 3851–7. doi:10.1007/s11033-012-2465-3. PMID 23271133.
33. Truninger, Kaspar; Menigatti, Mirco; Luz, Judith; Russell, Anna; Haider, Ritva; Gebbers, Jan-Olaf; Bannwart, Fridolin; Yurtsever, Hueseyin; Neuweiler, Joerg; Riehle, Hans-Martin; Cattaruzza, Maria Sofia; Heinimann, Karl; Schär, Primo; Jiricny, Josef; Marra, Giancarlo (2005). "Immunohistochemical Analysis Reveals High Frequency of PMS2 Defects in Colorectal Cancer". Gastroenterology. 128 (5): 1160–1171. doi:10.1053/j.gastro.2005.01.056. PMID 15887099.

34. Valeri, N.; Gasparini, P.; Fabbri, M.; Braconi, C.; Veronese, A.; Lovat, F.; Adair, B.; Vannini, I.; Fanini, F.; Bottoni, A.; Costinean, S.; Sandhu, S. K.; Nuovo, G. J.; Alder, H.; Gafa, R.; Calore, F.; Ferracin, M.; Lanza, G.; Volinia, S.; Negrini, M.; McIlhatton, M. A.; Amadori, D.; Fishel, R.; Croce, C. M. (2010). "Modulation of mismatch repair and genomic stability by miR-155". Proceedings of the National Academy of Sciences. 107 (15): 6982–6987. doi:10.1073/pnas.1002472107. PMC 2872463 . PMID 20351277.

35. Facista, Alexander; Nguyen, Huy; Lewis, Cristy; Prasad, Anil R; Ramsey, Lois; Zaitlin, Beryl; Nfonsam, Valentine; Krouse, Robert S; Bernstein, Harris; Payne, Claire M; Stern, Stephen; Oatman, Nicole; Banerjee, Bhaskar; Bernstein, Carol (2012). "Deficient expression of DNA repair enzymes in early progression to sporadic colon cancer". Genome Integrity. 3 (1): 3. doi:10.1186/2041-9414-3-3. PMC 3351028 . PMID 22494821.

36. Narayanan, L.; Fritzell, J. A.; Baker, S. M.; Liskay, R. M.; Glazer, P. M. (1997). "Elevated levels of mutation in multiple tissues of mice deficient in the DNA mismatch repair gene Pms2". Proceedings of the National Academy of Sciences. 94 (7): 3122–3127. doi:10.1073/pnas.94.7.3122. PMC 20332 . PMID 9096356.

37. Hegan, D. C.; Narayanan, L.; Jirik, F. R.; Edelmann, W.; Liskay, R.M.; Glazer, P. M. (2006). "Differing patterns of genetic instability in mice deficient in the mismatch repair genes Pms2, Mlh1, Msh2, Msh3 and Msh6". Carcinogenesis. 27 (12): 2402–2408. doi:10.1093/carcin/bgl079. PMC 2612936 . PMID 16728433.

38. Tutt, A. N.J.; Van Oostrom, CT; Ross, GM; Van Steeg, H; Ashworth, A (2002). "Disruption of Brca2 increases the spontaneous mutation rate in vivo: Synergism with ionizing radiation". EMBO Reports. 3 (3): 255–260. doi:10.1093/embo-reports/kvf037. PMC 1084010 . PMID 11850397.

39. German, J (1969). "Bloom's syndrome. I. Genetical and clinical observations in the first twenty-seven patients". American Journal of Human Genetics. 21 (2): 196–227. PMC 1706430 . PMID 5770175.

40. O'Hagan, Heather M.; Mohammad, Helai P.; Baylin, Stephen B. (2008). Lee, Jeannie T, ed. "Double Strand Breaks Can Initiate Gene Silencing and SIRT1-Dependent Onset of DNA Methylation in an Exogenous Promoter CpG Island". PLoS Genetics. 4 (8): e1000155. doi:10.1371/journal.pgen.1000155. PMC 2491723 . PMID 18704159.

41. Cuozzo, Concetta; Porcellini, Antonio; Angrisano, Tiziana; Morano, Annalisa; Lee, Bongyong; Di Pardo, Alba Di; Messina, Samantha; Iuliano, Rodolfo; Fusco, Alfredo; Santillo, Maria R.; Muller, Mark T.; Chiariotti, Lorenzo; Gottesman, Max E.; Avvedimento, Enrico V. (2007). "DNA Damage, Homology-Directed Repair, and DNA Methylation". PLoS Genetics. 3 (7): e110. doi:10.1371/journal.pgen.0030110. PMC 1913100 . PMID 17616978.

42. Roy, M; Sinha, D; Mukherjee, S; Biswas, J (2011). "Curcumin prevents DNA damage and enhances the repair potential in a chronically arsenic-exposed human population in West Bengal, India". European Journal of Cancer Prevention. 20 (2): 123–131. doi:10.1097/CEJ.0b013e328341017a. PMID 21332098.

43. Qin, XJ; Liu, W; Li, YN; Sun, X; Hai, CX; Hudson, LG; Liu, KJ (2012). "Poly(ADP-ribose) polymerase-1 inhibition by arsenite promotes the survival of cells with unrepaired DNA lesions induced by UV exposure". Toxicological Sciences. 127 (1): 120–129. doi:10.1093/toxsci/kfs099. PMC 3327874 . PMID 22387748.

44. Ebert, F; Weiss, A; Bültemeyer, M; Hamann, I; Hartwig, A; Schwerdtle, T (2011). "Arsenicals affect base excision repair by several mechanisms". Mutation research. 715 (1–2): 32–41. doi:10.1016/j.mrfmmm.2011.07.004. PMID 21782832.

45. Andrew, AS; Karagas, MR; Hamilton, JW (2003). "Decreased DNA repair gene expression among individuals exposed to arsenic in United States drinking water". International Journal of Cancer. Journal International Du Cancer. 104 (3): 263–268. doi:10.1002/ijc.10968. PMID 12569548.

46. Nollen, M; Ebert, F; Moser, J; Mullenders, LH; Hartwig, A; Schwerdtle, T (2009). "Impact of arsenic on nucleotide excision repair: XPC function, protein level, and gene expression". Molecular nutrition & food research. 53 (5): 572–582. doi:10.1002/mnfr.200800480. PMID 19382146.

47. Shen, S; Lee, J; Weinfeld, M; Le, XC (2008). "Attenuation of DNA damage-induced p53 expression by arsenic: A possible mechanism for arsenic co-carcinogenesis". Molecular carcinogenesis. 47 (7): 508–518. doi:10.1002/mc.20406. PMID 18085531.

48. Zhou, ZH; Lei, YX; Wang, CX (2012). "Analysis of aberrant methylation in DNA repair genes during malignant transformation of human bronchial epithelial cells induced by cadmium". Toxicological Sciences. 125 (2): 412–417. doi:10.1093/toxsci/kfr320. PMID 22112500.

49. Wieland, M; Levin, MK; Hingorani, KS; Biro, FN; Hingorani, MM (2009). "Mechanism of cadmium-mediated inhibition of Msh2-Msh6 function in DNA mismatch repair". Biochemistry. 48 (40): 9492–9502. doi:10.1021/bi9001248. PMID 19320425.

50. Bravard, A; Vacher, M; Gouget, B; Coutant, A; De Boisferon, FH; Marsin, S; Chevillard, S; Radicella, JP (2006). "Redox regulation of human OGG1 activity in response to cellular oxidative stress". Molecular and Cellular Biology. 26 (20): 7430–7436. doi:10.1128/MCB.00624-06. PMC 1636869 . PMID 16923968.

51. Viau, M; Gastaldo, J; Bencokova, Z; Joubert, A; Foray, N (2008). "Cadmium inhibits non-homologous end-joining and over-activates the MRE11-dependent repair pathway". Mutation research. 654 (1): 13–21. doi:10.1016/j.mrgentox.2008.04.009. PMID 18539077.

52. Schwerdtle, T; Ebert, F; Thuy, C; Richter, C; Mullenders, LH; Hartwig, A (2010). "Genotoxicity of soluble and particulate cadmium compounds: Impact on oxidative DNA damage and nucleotide excision repair". Chemical research in toxicology. 23 (2): 432–442. doi:10.1021/tx900444w. PMID 20092276.

53. Villeneuve, PJ; Mao Y (November 1994). "Lifetime probability of developing lung cancer, by smoking status, Canada". Canadian Journal of Public Health. 85 (6): 385–388. PMID 7895211.

54. Gerlinger, Marco; Rowan, Andrew J.; Horswell, Stuart; Larkin, James; Endesfelder, David; Gronroos, Eva; Martinez, Pierre; Matthews, Nicholas; Stewart, Aengus; Tarpey, Patrick; Varela, Ignacio; Phillimore, Benjamin; Begum, Sharmin; McDonald, Neil Q.; Butler, Adam; Jones, David; Raine, Keiran; Latimer, Calli; Santos, Claudio R.; Nohadani, Mahrokh; Eklund, Aron C.; Spencer-Dene, Bradley; Clark, Graham; Pickering, Lisa; Stamp, Gordon; Gore, Martin; Szallasi, Zoltan; Downward, Julian; Futreal, P. Andrew; Swanton, Charles (2012). "Intratumor Heterogeneity and Branched Evolution Revealed by Multiregion Sequencing". New England Journal of Medicine. 366 (10): 883–892. doi:10.1056/NEJMoa1113205. PMID 22397650.

55. Slaughter, Danely P.; Southwick, Harry W.; Smejkal, Walter (1953). ""Field cancerization" in oral stratified squamous epithelium. Clinical implications of multicentric origin". Cancer. 6 (5): 963–968. doi:10.1002/1097-0142(195309)6:5<963::AID-CNCR2820060515>3.0.CO;2-Q. PMID 13094644.

56. Bernstein, Carol; Bernstein, Harris; Payne, Claire M.; Dvorak, Katerina; Garewal, Harinder (2008). "Field defects in progression to gastrointestinal tract cancers". Cancer Letters. 260 (1–2): 1–10. doi:10.1016/j.canlet.2007.11.027. PMC 2744582 . PMID 18164807. Review

57. Nguyen, H; Loustaunau, C; Facista, A; Ramsey, L; Hassounah, N; Taylor, H; Krouse, R; Payne, CM; Tsikitis, VL; Goldschmid, S; Banerjee, B; Perini, RF; Bernstein, C (2010). "Deficient Pms2, ERCC1, Ku86, CcOI in field defects during progression to colon cancer". Journal of visualized experiments (41): 1931. doi:10.3791/1931. PMC 3149991 . PMID 20689513. The 28 minute video in this article is best watched by clicking on "Download video file" under the image in the abstract at https://www.ncbi.nlm.nih.gov/pmc/articles/PMC3149991/

58. Rubin, Harry (2011). "Fields and field cancerization: The preneoplastic origins of cancer". BioEssays. 33 (3): 224–231. doi:10.1002/bies.201000067. PMID 21254148.

59. Tsao, JL; Yatabe, Y; Salovaara, R; Järvinen, HJ; Mecklin, JP; Aaltonen, LA; Tavaré, S; Shibata, D (2000). "Genetic reconstruction of individual colorectal tumor histories". Proceedings of the National Academy of Sciences of the United States of America. 97 (3): 1236–1241. doi:10.1073/pnas.97.3.1236. PMC 15581 . PMID 10655514.

60. Vogelstein, B.; Papadopoulos, N.; Velculescu, V. E.; Zhou, S.; Diaz Jr, L. A.; Kinzler, K. W. (2013). "Cancer Genome Landscapes". Science. 339 (6127): 1546–58. doi:10.1126/science.1235122. PMC 3749880 . PMID 23539594. Review.

61. Shen, L.; Kondo, Y.; Rosner, G. L.; Xiao, L.; Hernandez, N. S.; Vilaythong, J.; Houlihan, P. S.; Krouse, R. S.; Prasad, A. R.; Einspahr, J. G.; Buckmeier, J.; Alberts, D. S.; Hamilton, S. R.; Issa, J.-P. J. (2005). "MGMT Promoter Methylation and Field Defect in Sporadic Colorectal Cancer". JNCI Journal of the National Cancer Institute. 97 (18): 1330–1338. doi:10.1093/jnci/dji275. PMID 16174854.

62. Svrcek, M.; Buhard, O.; Colas, C.; Coulet, F.; Dumont, S.; Massaoudi, I.; Lamri, A.; Hamelin, R.; Cosnes, J.; Oliveira, C.; Seruca, R.; Gaub, M.-P.; Legrain, M.;

Collura, A.; Lascols, O.; Tiret, E.; Fléjou, J.-F.; Duval, A. (2010). "Methylation tolerance due to an O6-methylguanine DNA methyltransferase (MGMT) field defect in the colonic mucosa: An initiating step in the development of mismatch repair-deficient colorectal cancers". Gut. 59 (11): 1516–26. doi:10.1136/gut.2009.194787. PMID 20947886.

63. Paluszczak, Jarosław; Misiak, Paulina; Wierzbicka, Małgorzata; Woźniak, Aldona; Baer-Dubowska, Wanda (2011). "Frequent hypermethylation of DAPK, RARbeta, MGMT, RASSF1A and FHIT in laryngeal squamous cell carcinomas and adjacent normal mucosa". Oral Oncology. 47 (2): 104–7. doi:10.1016/j.oraloncology.2010.11.006. PMID 21147548.
64. Zuo, Chunlai; Zhang, Haihong; Spencer, Horace J.; Vural, Emre; Suen, James Y.; Schichman, Steven A.; Smoller, Bruce R.; Kokoska, Mimi S.; Fan, Chun-Yang (2009). "Increased microsatellite instability and epigenetic inactivation of the hMLH1 gene in head and neck squamous cell carcinoma". Otolaryngology - Head and Neck Surgery. 141 (4): 484–90. doi:10.1016/j.otohns.2009.07.007. PMID 19786217.
65. Tawfik, Heba Mohamed; El-Maqsoud, Nehad M.R. Abd; Hak, Balegh H.A. Abdel; El-Sherbiny, Yasser M. (2011). "Head and neck squamous cell carcinoma: Mismatch repair immunohistochemistry and promoter hypermethylation of hMLH1 gene". American Journal of Otolaryngology. 32 (6): 528–36. doi:10.1016/j.amjoto.2010.11.005. PMID 21353335.
66. Zou, Xiao-Ping; Zhang, Bin; Zhang, Xiao-Qi; Chen, Min; Cao, Jun; Liu, Wen-Jia (2009). "Promoter hypermethylation of multiple genes in early gastric adenocarcinoma and precancerous lesions". Human Pathology. 40 (11): 1534–42. doi:10.1016/j.humpath.2009.01.029. PMID 19695681.
67. Wani, Majid; Afroze, Dil; Makhdoomi, Muzamil; Hamid, Iqra; Wani, Bilal; Bhat, Gulzar; Wani, Rauf; Wani, Khursheed (2012). "Promoter Methylation Status of DNA Repair Gene (hMLH1) in Gastric Carcinoma Patients of the Kashmir Valley". Asian Pacific Journal of Cancer Prevention. 13 (8): 4177–4181. doi:10.7314/APJCP.2012.13.8.4177. PMID 23098428.
68. Agarwal, A; Polineni, R; Hussein, Z; Vigoda, I; Bhagat, TD; Bhattacharyya, S; Maitra, A; Verma, A (2012). "Role of epigenetic alterations in the pathogenesis of Barrett's esophagus and esophageal adenocarcinoma". International Journal of Clinical and Experimental Pathology. 5 (5): 382–96. PMC 3396065 . PMID 22808291. Review.
69. Hofstad, B; Vatn, M H; Andersen, S N; Huitfeldt, H S; Rognum, T; Larsen, S; Osnes, M (1996). "Growth of colorectal polyps: Redetection and evaluation of unresected polyps for a period of three years". Gut. 39 (3): 449–456. doi:10.1136/gut.39.3.449. PMC 1383355 . PMID 8949653.
70. Schmitt, Michael W.; Prindle, Marc J.; Loeb, Lawrence A. (2012). "Implications of genetic heterogeneity in cancer". Annals of the New York Academy of Sciences. 1267: 110–6. doi:10.1111/j.1749-6632.2012.06590.x. PMC 3674777 . PMID 22954224.

71. Lander, Eric S.; Linton, Lauren M.; Birren, Bruce; Nusbaum, Chad; Zody, Michael C.; Baldwin, Jennifer; Devon, Keri; Dewar, Ken; Doyle, Michael; Fitzhugh, William; Funke, Roel; Gage, Diane; Harris, Katrina; Heaford, Andrew; Howland, John; Kann, Lisa; Lehoczky, Jessica; Levine, Rosie; McEwan, Paul; McKernan, Kevin; Meldrim, James; Mesirov, Jill P.; Miranda, Cher; Morris, William; Naylor, Jerome; Raymond, Christina; Rosetti, Mark; Santos, Ralph; Sheridan, Andrew; Sougnez, Carrie (2001). "Initial sequencing and analysis of the human genome". Nature. 409 (6822): 860–921. doi:10.1038/35057062. PMID 11237011.

72. Yost, S. E.; Smith, E. N.; Schwab, R. B.; Bao, L.; Jung, H.; Wang, X.; Voest, E.; Pierce, J. P.; Messer, K.; Parker, B. A.; Harismendy, O.; Frazer, K. A. (2012). "Identification of high-confidence somatic mutations in whole genome sequence of formalin-fixed breast cancer specimens". Nucleic Acids Research. 40 (14): e107. doi:10.1093/nar/gks299. PMC 3413110 . PMID 22492626.

73. Berger MF, Hodis E, Heffernan TP, Deribe YL, Lawrence MS, Protopopov A, Ivanova E, Watson IR, Nickerson E, Ghosh P, Zhang H, Zeid R, Ren X, Cibulskis K, Sivachenko AY, Wagle N, Sucker A, Sougnez C, Onofrio R, Ambrogio L, Auclair D, Fennell T, Carter SL, Drier Y, Stojanov P, Singer MA, Voet D, Jing R, Saksena G, Barretina J, Ramos AH, Pugh TJ, Stransky N, Parkin M, Winckler W, Mahan S, Ardlie K, Baldwin J, Wargo J, Schadendorf D, Meyerson M, Gabriel SB, Golub TR, Wagner SN, Lander ES, Getz G, Chin L, Garraway LA (2012). "Melanoma genome sequencing reveals frequent PREX2 mutations". Nature. 485 (7399): 502–6. doi:10.1038/nature11071. PMC 3367798 . PMID 22622578.

74. Rasnick, D; Duesberg, PH (1999). "How aneuploidy affects metabolic control and causes cancer". The Biochemical Journal. 340 (3): 621–30. doi:10.1042/0264-6021:3400621. PMC 1220292 . PMID 10359645.

75. López-Lázaro, M (2010). "A New View of Carcinogenesis and an Alternative Approach to Cancer Therapy". Molecular medicine. 16 (3–4): 144–153. doi:10.2119/molmed.2009.00162. PMC 2802554 . PMID 20062820.

76. A M Soto; C Sonnenschein (2004). "The somatic mutation theory of cancer: growing problems with the paradigm?". BioEssays. 26: 1097–1107. doi:10.1002/bies.20087.

77. P C W Davies; C H Lineweaver (February 2011). "Cancer tumors as Metazoa 1.0: tapping genes of ancient ancestors". Phys. Biol. 8 (1): 1–7. doi:10.1088/1478-3975/8/1/015001.

78. Dean, Tim. "Cancer resembles life 1 billion years ago, say astrobiologists", Australian Life Scientist, 8 February 2011. Retrieved 15 February 2011.

79. Nowell PC (October 1976). "The clonal evolution of tumor cell populations". Science. 194 (4260): 23–8. doi:10.1126/science.959840. PMID 959840.

80. Merlo LM, Pepper JW, Reid BJ, Maley CC (December 2006). "Cancer as an evolutionary and ecological process". Nature Reviews Cancer. 6 (12): 924–35. doi:10.1038/nrc2013. PMID 17109012.

81. Hanahan D, Weinberg RA; Weinberg (2000). "The hallmarks of cancer". Cell.

100 (1): 57–70. doi:10.1016/S0092-8674(00)81683-9. PMID 10647931.

82. Cho RW, Clarke MF; Clarke (February 2008). "Recent advances in cancer stem cells". Current Opinion in Genetics & Development. 18 (1): 48–53. doi:10.1016/j.gde.2008.01.017. PMID 18356041.
83. Taniguchi K, Wu LW, Grivennikov SI, de Jong PR, Lian I, Yu FX, Wang K, Ho SB, Boland BS, Chang JT, Sandborn WJ, Hardiman G, Raz E, Maehara Y, Yoshimura A, Zucman-Rossi J, Guan KL, Karin M (Mar 2015). "A gp130-Src-YAP module links inflammation to epithelial regeneration". Nature. 519 (7541): 57–62. doi:10.1038/nature14228. PMC 4447318 . PMID 25731159.
84. You H, Lei P, Andreadis ST (Dec 2013). "JNK is a novel regulator of intercellular adhesion". Tissue Barriers. 1 (5): e26845. doi:10.4161/tisb.26845. PMC 3942331 . PMID 24868495.
85. Busillo JM, Azzam KM, Cidlowski JA (Nov 2011). "Glucocorticoids sensitize the innate immune system through regulation of the NLRP3 inflammasome". J Biol Chem. 286 (44): 38703–13. doi:10.1074/jbc.M111.275370. PMC 3207479 . PMID 21940629.
86. Wang Y, Bugatti M, Ulland TK, Vermi W, Gilfillan S, Colonna M (Dec 2015). "Non-redundant roles of keratinocyte-derived IL-34 and neutrophil-derived CSF1 in Langerhans cells renewal in the steady-state and skin inflammation". Eur J Immunol. doi:10.1002/eji.201545917. PMID 26634935.
87. Siqueira Mietto B, Kroner A, Girolami EI, Santos-Nogueira E, Zhang J, David S (Dec 2015). "Role of IL-10 in Resolution of Inflammation and Functional Recovery after Peripheral Nerve Injury". J Neurosci. 35 (50): 16431–42. doi:10.1523/JNEUROSCI.2119-15.2015. PMID 26674868.
88. Seifert AW, Maden M (2014). "New insights into vertebrate skin regeneration". Int Rev Cell Mol Biol. 310: 129–69. doi:10.1016/B978-0-12-800180-6.00004-9. PMID 24725426.
89. Kwon MJ, Shin HY, Cui Y, Kim H, Thi AH, Choi JY, Kim EY, Hwang DH, Kim BG (Dec 2015). "CCL2 Mediates Neuron-Macrophage Interactions to Drive Proregenerative Macrophage Activation Following Preconditioning Injury". J Neurosci. 35 (48): 15934–47. doi:10.1523/JNEUROSCI.1924-15.2015. PMID 26631474.
90. Hajishengallis G, Chavakis T (Jan 2013). "Endogenous modulators of inflammatory cell recruitment". Trends in Immunology. 34 (1): 1–6. doi:10.1016/j.it.2012.08.003. PMID 22951309.
91. Nelson AM, Katseff AS, Ratliff TS, Garza LA (Nov 2015). "Interleukin 6 and STAT3 regulate p63 isoform expression in keratinocytes during regeneration". Exp Dermatol. 25: 155–7. doi:10.1111/exd.12896. PMID 26566817.
92. Vidal PM, Lemmens E, Dooley D, Hendrix S (Feb 2013). "The role of "anti-inflammatory" cytokines in axon regeneration". Cytokine & Growth Factor Reviews. 24 (1): 1–12. doi:10.1016/j.cytogfr.2012.08.008. PMID 22985997.
93. Hsueh YY, Chang YJ, Huang CW, Handayani F, Chiang YL, Fan SC, Ho CJ, Kuo YM, Yang SH, Chen YL, Lin SC, Huang CC, Wu CC (Oct 2015). "Synergy

of endothelial and neural progenitor cells from adipose-derived stem cells to preserve neurovascular structures in rat hypoxic-ischemic brain injury". Sci Rep. 5: 14985. doi:10.1038/srep14985. PMC 4597209 . PMID 26447335.

94. Yaniv M (Sep 2014). "Chromatin remodeling: from transcription to cancer". Cancer Genetics. 207 (9): 352–7. doi:10.1016/j.cancergen.2014.03.006. PMID 24825771.

95. Zhang X, He N, Gu D, Wickliffe J, Salazar J, Boldogh I, Xie J (Oct 2015). "Genetic Evidence for XPC-KRAS Interactions During Lung Cancer Development". J Genet Genomics. 42 (10): 589–96. doi:10.1016/j.jgg.2015.09.006. PMID 26554912.

96. Dubois-Pot-Schneider H, Fekir K, Coulouarn C, Glaise D, Aninat C, Jarnouen K, Le Guével R, Kubo T, Ishida S, Morel F, Corlu A (Dec 2014). "Inflammatory cytokines promote the retrodifferentiation of tumor-derived hepatocyte-like cells to progenitor cells". Hepatology. 60 (6): 2077–90. doi:10.1002/hep.27353. PMID 25098666.

97. Finkin S, Yuan D, Stein I, Taniguchi K, Weber A, Unger K, Browning JL, Goossens N, Nakagawa S, Gunasekaran G, Schwartz ME, Kobayashi M, Kumada H, Berger M, Pappo O, Rajewsky K, Hoshida Y, Karin M, Heikenwalder M, Ben-Neriah Y, Pikarsky E (Dec 2015). "Ectopic lymphoid structures function as microniches for tumor progenitor cells in hepatocellular carcinoma". Nat Immunol. 16 (12): 1235–44. doi:10.1038/ni.3290. PMID 26502405.

98. Vlahopoulos SA, Cen O, Hengen N, Agan J, Moschovi M, Critselis E, Adamaki M, Bacopoulou F, Copland JA, Boldogh I, Karin M, Chrousos GP (Aug 2015). "Dynamic aberrant NF-κB spurs tumorigenesis: a new model encompassing the microenvironment". Cytokine & Growth Factor Reviews. 26 (4): 389–403. doi:10.1016/j.cytogfr.2015.06.001. PMID 26119834.

99. Grivennikov SI, Karin M (Feb 2010). "Dangerous liaisons: STAT3 and NF-kappaB collaboration and crosstalk in cancer". Cytokine & Growth Factor Reviews. 21 (1): 11–9. doi:10.1016/j.cytogfr.2009.11.005. PMID 20018552.

100. Rieger S, Zhao H, Martin P, Abe K, Lisse TS (Jan 2015). "The role of nuclear hormone receptors in cutaneous wound repair". Cell Biochem Funct. 33 (1): 1–13. doi:10.1002/cbf.3086. PMID 25529612.

101. Lu X, Yarbrough WG (Feb 2015). "Negative regulation of RelA phosphorylation: emerging players and their roles in cancer". Cytokine & Growth Factor Reviews. 26 (1): 7–13. doi:10.1016/j.cytogfr.2014.09.003. PMID 25438737.

102. Vlahopoulos SA, Cen O, Hengen N, Agan J, Moschovi M, Critselis E, Adamaki M, Bacopoulou F, Copland JA, Boldogh I, Karin M, Chrousos GP (Aug 2015). "Dynamic aberrant NF-κB spurs tumorigenesis: a new model encompassing the microenvironment". Cytokine & Growth Factor Reviews. 26 (4): 389–403. doi:10.1016/j.cytogfr.2015.06.001. PMID 26119834.

103. Sionov RV, Fridlender ZG, Granot Z (Jun 2014). "The Multifaceted Roles Neutrophils Play in the Tumor Microenvironment". Cancer Microenvironment. doi:10.1007/s12307-014-0147-5. PMID 24895166.

104. Venturi, Sebastiano (2011). "Evolutionary Significance of Iodine". Current

Chemical Biology-. 5 (3): 155–162. doi:10.2174/187231311796765012. ISSN 1872-3136.

105. Venturi, Sebastiano (2014). "Iodine, PUFAs and Iodolipids in Health and Disease: An Evolutionary Perspective". Human Evolution-. 29 (1-3): 185–205. ISSN 0393-9375.
106. Walsh CJ, Luer CA, Bodine AB, Smith CA, Cox HL, Noyes DR, Maura G (2006). "Elasmobranch immune cells as a source of novel tumor cell inhibitors: Implications for public health.". Integr Comp Biol. 46 (6): 1072–1081. doi:10.1093/icb/icl041. PMC 2664222 . PMID 19343108.
107. Vogelstein, Bert; Kinzler, Kenneth W (2004). "Cancer genes and the pathways they control". Nature Medicine. 10 (8): 789–99. doi:10.1038/nm1087. PMID 15286780.
108. Brand, KA; Hermfisse, U (1997). "Aerobic glycolysis by proliferating cells: a protective strategy against reactive oxygen species". The FASEB Journal. 11 (5): 388–95. PMID 9141507.
109. Dingcheng Gao, Nolan DJ, Mellick AS, et al. (January 11, 2008). "Endothelial Progenitor Cells Control the Angiogenic Switch in Mouse Lung Metastasis". Science. Washington, D.C.: American Association for the Advancement of Science. 319 (5860): 195–198. doi:10.1126/science.1150224. ISSN 0036-8075. PMID 18187653.
110. Nolan DJ, Ciarrocchi A, Mellick AS, et al. (June 15, 2007). "Bone marrow-derived endothelial progenitor cells are a major determinant of nascent tumor neovascularization". Genes & Development. Cold Spring Harbor, NY: Cold Spring Harbor Laboratory Press. 21 (12): 1546–1558. doi:10.1101/gad.436307. ISSN 0890-9369. PMC 1891431 . PMID 17575055.
111. Mellick As; Plummer PN; et al. (2010). "Using the Transcription Factor Inhibitor of DNA Binding 1 to Selectively Target Endothelial Progenitor Cells Offers Novel Strategies to Inhibit Tumor Angiogenesis and Growth". Cancer Research. 70 (18): 7273–7282. doi:10.1158/0008-5472.CAN-10-1142. PMC 3058751 . PMID 20807818.
112. Bos JL (September 1989). "ras oncogenes in human cancer: a review". Cancer Research. 49 (17): 4682–9. PMID 2547513. Retrieved 6 June 2009.
113. Chang EH, Furth ME, Scolnick EM, Lowy DR (1982). "Tumorigenic transformation of mammalian cells induced by a normal human gene homologous to the oncogene of Harvey murine sarcoma virus". Nature. 297 (5866): 479–83. doi:10.1038/297479a0. PMID 6283358.
114. Vlahopoulos SA, Logotheti S, Mikas D, Giarika A, Gorgoulis V, Zoumpourlis V (April 2008). "The role of ATF-2 in oncogenesis". BioEssays. 30 (4): 314–27. doi:10.1002/bies.20734. PMID 18348191.
115. Matoba S, Kang J, Patino W, Wragg A, Boehm M, Gavrilova O, Hurley P, Bunz F, Hwang P (2006). "p53 regulates mitochondrial respiration". Science. 312 (5780): 1650–3. doi:10.1126/science.1126863. PMID 16728594.
116. Knudson A (1971). "Mutation and Cancer: Statistical Study of Retinoblastoma".

Proc Natl Acad Sci USA. 68 (4): 820–3. doi:10.1073/pnas.68.4.820. PMC 389051 . PMID 5279523.

117. Fodde R, Smits R; Smits (2002). "Cancer biology. A matter of dosage". Science. 298 (5594): 761–3. doi:10.1126/science.1077707. PMID 12399571.

118. Stephens, Philip J.; Greenman, Chris D.; Fu, Beiyuan; et al. (January 2011). "Massive Genomic Rearrangement Acquired in a Single Catastrophic Event during Cancer Development". Cell. 144 (1): 27–40. doi:10.1016/j.cell.2010.11.055. PMC 3065307 . PMID 21215367. Lay summary – The New York times (10 January 2011).

119. Carrillo-Infante C, Abbadessa G, Bagella L, Giordano A (June 2007). "Viral infections as a cause of cancer (review)". Int. J. Oncol. 30 (6): 1521–8. doi:10.3892/ijo.30.6.1521. PMID 17487374.

120. Safdar, Amar (2011-06-01). Management of Infections in Cancer Patients. Springer. pp. 478–. ISBN 978-1-60761-643-6. Retrieved 17 August 2011.

121. Samaras V, Rafailidis PI, Mourtzoukou EG, Peppas G, Falagas ME (May 2010). "Chronic bacterial and parasitic infections and cancer: a review". J Infect Dev Ctries. 4 (5): 267–81. doi:10.3855/jidc.819. PMID 20539059.

122. Dalerba, P.; Cho, R. W.; Clarke, M. F. (2007). "Cancer stem cells: models and concepts". Annu. Rev. Med. 58: 267–284. doi:10.1146/annurev.med.58.062105.204854. PMID 17002552.

123. Nowell PC (October 1976). "The clonal evolution of tumor cell populations". Science. 194 (4260): 23–8. doi:10.1126/science.959840. PMID 959840.

124. Zhang W, Hanks AN, Boucher K, et al. (January 2005). "UVB-induced apoptosis drives clonal expansion during skin tumor development". Carcinogenesis. 26 (1): 249–57. doi:10.1093/carcin/bgh300. PMC 2292404 . PMID 15498793.

125. Belostotsky. "Mechanical trigger of canrcinogenesis". Schemes Journal.

Index

P

R

S

T

V